Synthesis Lectures on Engineering, Science, and Technology

The focus of this series is general topics, and applications about, and for, engineers and scientists on a wide array of applications, methods and advances. Most titles cover subjects such as professional development, education, and study skills, as well as basic introductory undergraduate material and other topics appropriate for a broader and less technical audience.

Navid Asadizanjani ·
Himanandhan Reddy Kottur · Hamed Dalir

Introduction to Microelectronics Advanced Packaging Assurance

Navid Asadizanjani
University of Florida
Gainesville, FL, USA

Hamed Dalir
University of Florida
Gainesville, FL, USA

Himanandhan Reddy Kottur
University of Florida
Gainesville, FL, USA

ISSN 2690-0300   ISSN 2690-0327   (electronic)
Synthesis Lectures on Engineering, Science, and Technology
ISBN 978-3-031-86101-7   ISBN 978-3-031-86102-4   (eBook)
https://doi.org/10.1007/978-3-031-86102-4

This Springer imprint is published by the registered company Springer Nature Switzerland AG
The registered company address is: Gewerbestrasse 11, 6330 Cham, Switzerland

If disposing of this product, please recycle the paper.

To our families and friends, for your endless love and encouragement. Your support has been invaluable in this journey, and we are profoundly grateful for the sacrifices you've made to help us pursue our dreams.

Preface

The field of semiconductor packaging has evolved beyond simply protecting integrated circuits (ICs) to become a critical determinant of device performance, reliability, and security. As electronics grow smaller, faster, and more interconnected, packaging techniques have progressed from traditional methods like wire bonding to advanced solutions such as flip-chip bonding, through-silicon vias (TSVs), and hybrid bonding. Alongside these technical advancements, there is a growing focus on supply chain security, quality assurance, and sustainability.

This book, *Introduction to Microelectronics Advanced Packaging Assurance*, offers both a technical guide and strategic roadmap for engineers, researchers, and professionals interested in advanced packaging. It emphasizes the importance of security throughout the lifecycle—from design to manufacturing—addressing issues such as counterfeit detection and blockchain-based traceability. With increased reliance on semiconductors across critical sectors like healthcare, defense, and communications, the integrity of the entire packaging process is more important than ever.

Chapter 1 explores the evolution of IC packaging and assembly processes, setting the foundation for understanding the industry's shift from traditional to modern methods. Chapter 2 delves into bonding interconnects, covering essential techniques such as wire bonding, flip-chip, micro bumps, and TSVs, examining their principles, applications, and limitations. Additional chapters focus on Physical Vapor Deposition (PVD), Chemical Vapor Deposition (CVD), and etching techniques, detailing their role in achieving high-performance packaging.

In its penultimate chapter, the book addresses the security and assurance aspects of packaging, discussing strategies for monitoring supply chain integrity, tamper-evident techniques, and the use of AI in counterfeit detection. The final chapter concludes with Chemical Mechanical Planarization (CMP) and wafer grinding, providing insight into their importance in achieving smooth surfaces critical for advanced bonding techniques.

This book is written for both students and professionals. It offers accessible content for learners new to packaging while providing detailed insights for experienced practitioners, covering fabrication processes, materials, and the latest industry practices.

We hope *Introduction to Microelectronics Advanced Packaging Assurance* serves as a valuable resource, bridging the gap between theory and practice while highlighting the importance of security and innovation in microelectronics packaging.

Gainesville, USA　　　　　　　　　　　　　　　　　Navid Asadizanjani
October 2024　　　　　　　　　　　　　　Himanandhan Reddy Kottur
　　　　　　　　　　　　　　　　　　　　　　　　　Hamed Dalir

Acknowledgements

A heartfelt thank you to *Pavanbabu Arjunamahanti* for your invaluable support throughout the development of this book. Your contributions to various chapters, insights, and dedication have been instrumental in shaping the content and ensuring its depth and clarity. Your collaboration and encouragement were crucial at every step of the writing process, and for that, I am deeply grateful.

I would also like to extend my sincere thanks to *Liton Kumar Biswas, M. Shafkat M. Khan, and Nitin Varshney* for their continuous support and motivation throughout this journey.

Contents

1   **Introduction to IC Packaging Development and Assembly Processes**   1
    1.1   A Historical Perspective: From Simple Beginnings to Complex
        Systems   1
    1.2   Navigating Challenges in IC Packaging   3
    1.3   The Path to Packaging: Key Steps in IC Assembly   5
        1.3.1   Steps Involved in IC Packaging   7
    1.4   The Complex Ecosystem of Semiconductor Packaging   9
        1.4.1   The US Semiconductor Packaging Ecosystem: Challenges
            and Specific Metrics   11
        1.4.2   Domestic Expansion of Advanced Packaging Capabilities   13
    References   17

2   **Bonding Techniques–Interconnects**   21
    2.1   Role of Bonding in Packaging   21
    2.2   Wire Bonding: A Longstanding and Evolving Technology
        (1950-Present)   21
        2.2.1   Ball Bonding   23
        2.2.2   Wedge Bonding   24
        2.2.3   Addressing Signal Integrity and High-Frequency
            Challenges   25
    2.3   Flip Chip Bonding: Evolution and Advantages Over Wire Bonding   25
    2.4   Micro Bump (C2) and Solder Bump (C4): Evolution
        and Applications (1980-Present)   26
    2.5   TSV Bonding: Revolutionizing 3D Integration (Early 2000-Present) ...   27
    2.6   Thermocompression Bonding: Precision Through Heat
        and Pressure   30
    2.7   Thermosonic Bonding: Ultrasonic Energy for Delicate Components ...   31
    2.8   Anisotropic Conductive Film (ACF) Bonding   31
    2.9   Hybrid Bonding: The Backbone of 3D ICs and Heterogeneous
        Integration (2020-Present)   32

2.10   Materials for Bonding Interconnects   33
2.11   Challenges in Bonding Techniques   34
2.12   Industry Leaders and the Bonding Ecosystem   35
2.13   Equipment Vendors: Supporting Bonding Processes
       in High-Volume Production and R&D   36
References   38

3   CVD in Semiconductor Packaging   41
3.1   An Introduction to the World of Chemical Vapor Deposition   41
3.2   Principles of CVD   42
       3.2.1   Basic Mechanism   42
       3.2.2   Key Parameters   43
3.3   Types of CVD Processes   44
       3.3.1   Low-Pressure CVD (LPCVD)   44
       3.3.2   Plasma-Enhanced CVD (PECVD)   45
       3.3.3   Metal-Organic CVD (MOCVD)   45
       3.3.4   Atomic Layer Deposition (ALD)   45
       3.3.5   Sol-Gel CVD   47
       3.3.6   Hot-Wall CVD   48
       3.3.7   Cold-Wall CVD   48
3.4   CVD Materials in Semiconductor Packaging   48
       3.4.1   Dielectrics   48
       3.4.2   Conductors   49
       3.4.3   Semiconductors   50
3.5   Applications of CVD in Packaging   51
       3.5.1   CVD in TSVs   51
       3.5.2   CVD in RDLs   51
       3.5.3   CVD in Protective Coatings   51
3.6   The Broad Reach of CVD: Enhancing NEMS, Photonics,
       and Encapsulation Technologies   52
       3.6.1   CVD in MEMS and NEMS Fabrication   52
       3.6.2   CVD for Encapsulation Layer Deposition   52
       3.6.3   CVD in Silicon Photonics   53
3.7   CVD-Based Solutions for Critical Vulnerabilities   53
       3.7.1   Thermal Stress   53
       3.7.2   Chemical and Moisture Ingress   54
       3.7.3   Mechanical Stress and Fracture   54
       3.7.4   Electromigration   54
       3.7.5   Electrical Leakage and Crosstalk   55
       3.7.6   Contamination During Fabrication   55

3.8 Future Trends and Innovations in CVD Technology 55
    3.8.1 Nanostructured Materials 55
    3.8.2 Eco-Friendly CVD Processes Environmental Impact
          Reduction 56
3.9 Strategic Insights Into the CVD Technology Ecosystem 57
    3.9.1 Collaborative Innovations and Sector Growth 57
    3.9.2 Market Dynamics and Future Prospects 57
References 58

4 **Etching Techniques and Applications in Advanced IC Packaging** 61
4.1 Introduction 61
4.2 Classification of Etching Techniques 62
    4.2.1 Wet Etching 63
    4.2.2 Mechanism of Wet Etching 63
    4.2.3 Advantages of Wet Etching 63
    4.2.4 Challenges of Wet Etching 63
    4.2.5 Common Applications of Wet Etching 64
4.3 Dry Etching 64
    4.3.1 Mechanism of Dry Etching 64
    4.3.2 Types of Dry Etching 65
    4.3.3 Advantages of Dry Etching 65
    4.3.4 Challenges of Dry Etching 65
4.4 Process Parameters for High Precision Etching 66
    4.4.1 Etch Selectivity 66
    4.4.2 Etch Rate 67
    4.4.3 Anisotropy 68
    4.4.4 Isotropy 68
4.5 Etching Across Packaging Processes: From RDLs to TSVs 69
    4.5.1 RDL Patterning 70
    4.5.2 Through-Silicon Vias (TSVs) Formation 70
    4.5.3 Solder Bump Formation and Pad Preparation 70
    4.5.4 Polyimide Patterning for Flexible Packaging 71
    4.5.5 Passivation and Encapsulation Layer Etching 71
    4.5.6 Etching in MEMS Packaging 71
4.6 Etching-Based Solutions for Critical Vulnerabilities
    in Microelectronics 72
4.7 Future Trends in Etching 73
4.8 Industry Vendors and the Global Etching Ecosystem 74
    4.8.1 Lam Research 74
    4.8.2 Applied Materials 75
    4.8.3 Tokyo Electron Limited (TEL) 75
    4.8.4 ASM International 76

        4.8.5   Hitachi High-Technologies   76
        4.8.6   Plasma-Therm   76
        4.8.7   Oxford Instruments   77
        4.8.8   SPTS Technologies (an Orbotech Company)   77
    References   78

5   Physical Vapor Deposition in Advanced Semiconductor Packaging   81
    5.1   Introduction to Physical Vapor Deposition (PVD)   81
        5.1.1   Definition of PVD   81
        5.1.2   Historical Background and Evolution of PVD   81
    5.2   Types of PVD Processes   82
        5.2.1   Sputtering   83
    5.3   Evaporation   86
        5.3.1   E-Beam Evaporation   86
        5.3.2   Resistive Evaporation   86
        5.3.3   Arc Evaporation   87
        5.3.4   Inductive Thermal Evaporation   87
    5.4   Process Steps in PVD   88
        5.4.1   Substrate Preparation and Cleaning   88
        5.4.2   Loading Into the PVD Chamber   88
        5.4.3   Vacuum Creation and Process Environment Setup   88
        5.4.4   Film Deposition: Monitoring Thickness and Uniformity   89
        5.4.5   Cooling and Unloading   89
    5.5   Materials Deposited by PVD   89
        5.5.1   Dielectrics   90
        5.5.2   Transparent Conductive Oxides (TCOs): Indium Tin
                 Oxide (ITO)   90
        5.5.3   Alloys and Compounds: Titanium Nitride (TiN), Tantalum
                 Nitride (TaN)   90
    5.6   Applications of PVD in Semiconductor Manufacturing   91
        5.6.1   Interconnect Formation   91
        5.6.2   Barrier Layers for Diffusion Control (e.g., TiN, TaN)   91
        5.6.3   Contact Metallization for MEMS and Power Devices   92
        5.6.4   Thin-Film Resistors and Capacitors   92
        5.6.5   Decorative Coatings in Consumer Electronics   93
    5.7   PVD in Advanced Packaging   93
        5.7.1   RDLs for Wafer-Level Packaging (WLP)   93
        5.7.2   Through-Silicon Vias (TSVs) and Vertical Interconnects   94
        5.7.3   Wearable Electronics and Flexible Packaging Applications   95

5.8   Advantages and Challenges of PVD 96
    5.8.1   Advantages of PVD 96
    5.8.2   Challenges of PVD 97
5.9   PVD-Based Solutions for Solving Critical Vulnerabilities 98
    5.9.1   Electromigration Mitigation for Interconnects 98
    5.9.2   Diffusion Control with Barrier Layers 98
    5.9.3   Enhanced Adhesion to Prevent Delamination 99
    5.9.4   Surface Contamination Removal and Protection 99
    5.9.5   Signal Integrity Improvement for High-Frequency
             Applications 99
    5.9.6   Stress Management in Thick Films for MEMS and Power
             Devices 100
    5.9.7   Mechanical Flexibility for Wearable and Flexible
             Electronics 100
    5.9.8   Prevention of Corrosion and Environmental Degradation 100
    5.9.9   Thermal Management for High-Performance Devices 101
5.10   Emerging Trends and Innovations in PVD 101
5.11   PVD Ecosystem 102
    5.11.1   Groups of PVD Equipment and Component Suppliers 103
References 104

6   **The Crucial Role of CMP and Wafer Grinding** 107
6.1   Introduction to CMP and Wafer Grinding 107
6.2   Definition of CMP and Wafer Grinding 107
6.3   Historical Evolution in Semiconductor Manufacturing 108
6.4   Importance of CMP and Wafer Grinding in Achieving Flat,
    Defect-Free Surfaces 109
6.5   Role of CMP and Wafer Grinding in Modern Fabrication 110
6.6   Overview of CMP Process 110
    6.6.1   Working Principle 110
    6.6.2   CMP Consumables: Pads, Slurries, and Conditioners 111
    6.6.3   Key Steps of CMP Process 111
6.7   Overview of Wafer Grinding Process 112
    6.7.1   Working Principle 112
    6.7.2   Types of Grinding Processes 112
6.8   Fine Grinding Versus Rough Grinding: Surface Finish Control 113
6.9   CMP Materials and Consumables 114
6.10   Role of CMP and Wafer Grinding in Advanced Packaging
    Technologies 115
    6.10.1   CMP in RDLs: Planarizing Metal Layers for Wafer-Level
             Packaging 115

        6.10.2   CMP and TSV Processing: Ensuring Flat Surfaces
                 for TSVs in 3D Packaging 115
        6.10.3   Wafer Grinding for Flexible Electronics: Creating
                 Ultra-Thin Wafers for Wearable Devices 116
        6.10.4   Importance of Surface Finish Control in Chiplet
                 and Heterogeneous Integration 116
   6.11 CMP-Based Solutions for Solving Vulnerabilities 117
        6.11.1   Preventing Surface Defects and Contamination 117
        6.11.2   Addressing Thermal Stress and Mechanical Fatigue 117
        6.11.3   Reducing Power Loss and Parasitic Effects
                 in High-Density Interconnects 117
   6.12 Main Vendors in CMP and Wafer Grinding Technologies 118
        6.12.1   CMP Vendors 118
        6.12.2   Wafer Grinding Vendors 118
   References 119

7 **Electrochemical Deposition in Advanced Packaging** 121
   7.1   Introduction to Electrochemical Deposition (ECD) 121
   7.2   Definition of Electrochemical Deposition (ECD) 121
   7.3   Historical Development and Evolution of ECD in Semiconductor
         Manufacturing 122
   7.4   Types of Electrochemical Deposition Processes 123
        7.4.1   Electroplating 124
        7.4.2   Electroless Deposition 125
   7.5   Process Steps in Electrochemical Deposition (ECD) 126
   7.6   Materials Used in Electrochemical Deposition (ECD) 127
   7.7   Applications of ECD in Semiconductor Manufacturing 129
        7.7.1   Copper Interconnects: Supporting Signal Integrity
                 in Complex Circuits 129
        7.7.2   Wafer Bumping: Enabling Heterogeneous Integration
                 Through Advanced Microbumping 129
        7.7.3   Redistribution Layers (RDLs): Bridging the Gap Between
                 Chips and External Interfaces 129
        7.7.4   Thin-Film Deposition for MEMS: Functional Coatings
                 for Enhanced Device Reliability 130
   7.8   ECD-Based Solutions for Addressing Vulnerabilities 130
        7.8.1   Mitigating Electromigration in Copper Interconnects
                 Through Microstructural Control 130
        7.8.2   Achieving Void-Free Deposits with Advanced Plating
                 Techniques 130
        7.8.3   Reducing Stress-Induced Cracking with Multi-layer
                 Plating and Additives 131

| | 7.8.4 | Improving Adhesion and Reliability Through Seed Layers and Annealing | 131 |

7.8.4   Improving Adhesion and Reliability Through Seed Layers
        and Annealing 131
7.8.5   Corrosion Protection with Electroplated Gold and Nickel
        Layers 131
7.9   Challenges and Process Optimization in ECD 132
7.9.1   Void Formation and Poor Coverage in High-Aspect-Ratio
        Features 132
7.9.2   Controlling Film Stress and Grain Size for Reliability 132
7.9.3   Uniformity Issues Across Large Wafers: A Scaling
        Challenge 133
7.9.4   Process Optimization Techniques: Pulse Plating,
        Additives, and Dynamic Bath Control 133
7.10   Emerging Trends and Innovations in ECD 134
7.10.1   Pulse-Reverse Electroplating: Achieving Superior
         Uniformity 134
7.10.2   Hybrid ECD Processes: Combining Electroplating
         with Electroless Deposition 134
7.10.3   Precision Mask Jet ECD Process Flow 134
7.10.4   ECD for Advanced Interconnects in AI and 5G
         Applications 136
7.10.5   Sustainable ECD Processes: Green Chemistry
         and Recycling of Electrolytes 136
7.11   Industry Vendors and Ecosystem 137
7.11.1   Key Players in the ECD Ecosystem 137
7.11.2   Global Market Trends: Adoption in Advanced Packaging
         and Heterogeneous Integration 138
References 139

8 Testing and Reliability in Advanced Packaging 141
8.1   Introduction to Testing and Reliability in Packaging 141
8.2   Types of Packaging Reliability Tests 142
8.2.1   Mechanical Testing 142
8.2.2   Environmental Testing 143
8.2.3   Electrical Testing 144
8.2.4   Chemical Testing 144
8.3   Common Reliability Issues in Packaging 145
8.3.1   Cracking, Delamination, and Warping 145
8.3.2   Corrosion and Oxidation 146
8.3.3   Thermal Stress and Electromigration 147
8.3.4   Mechanical Failures 148

8.4   Non-destructive Testing (NDT) Methods 149
8.5   Destructive Testing Methods 150
8.6   Standards and Guidelines for Reliability Testing 152
        8.6.1   JEDEC Standards 152
        8.6.2   MIL-STD 153
        8.6.3   IPC Standards 153
        8.6.4   Compliance Requirements and Certifications 154
8.7   Reliability Modeling and Prediction 154
        8.7.1   Arrhenius Model 154
        8.7.2   Coffin-Manson Model 155
        8.7.3   MTBF (Mean Time Between Failures) 155
        8.7.4   FIT (Failures in Time) Rate 156
        8.7.5   Accelerated Life Testing 156
8.8   Future Trends in Packaging Reliability and Testing 156
        8.8.1   AI-Powered Testing Systems 157
        8.8.2   Real-Time Monitoring and Digital Twins 157
References 158

9 Quantum Computing, Wearables, and Next-Gen IC Packaging 161
9.1   Foundations of the Quantum and Wearable Revolution 161
9.2   Quantum Computing and Advanced Packaging Technologies 162
        9.2.1   The Rise of Quantum Computing 162
        9.2.2   3D Integration for Quantum Chips 163
        9.2.3   TSVs: Unlocking High-Performance Quantum Operations 165
        9.2.4   Thermal Management in 3D Quantum Chips 165
        9.2.5   Benefits of 3D Integration for Quantum Error Correction 166
        9.2.6   Intel's 3D Quantum Prototypes: A Case Study 166
9.3   Advanced Packaging for Wearable Devices 167
        9.3.1   Key Requirements for Wearables: Compact,
                Biocompatible, and Energy-Efficient 167
        9.3.2   Flexible and Bendable Packaging for Wearables 167
        9.3.3   Energy Harvesting and Self-Powered Systems 168
        9.3.4   Encapsulation for Rugged Wearables 169
9.4   Interposerless Technologies: The Next Wave in IC Packaging 170
        9.4.1   What is Interposerless Packaging? 170
        9.4.2   Applications of Interposerless Packaging 170
        9.4.3   Impact on AI Accelerators and 5G Networks 171
9.5   Advantages and Future Trends 172

9.6   Quantum Computing: Overcoming Decoherence and Ensuring
        Fault Tolerance 172
        9.6.1   Quantum Decoherence: Preserving Qubit Stability 172
        9.6.2   Fault Tolerance: Managing Errors in Quantum Operations 173
9.7   Wearable Devices: Ensuring Biocompatibility and Environmental
        Reliability 174
        9.7.1   Biocompatibility: Safe Interaction with the Human Body 174
        9.7.2   Environmental Reliability: Surviving Harsh Conditions 174
9.8   Research Directions and Future Opportunities 175
        9.8.1   Quantum Computing Research Directions 175
        9.8.2   Wearables Research Directions 176
References 178

Glossary 181

Introduction to IC Packaging Development and Assembly Processes

1

1.1 A Historical Perspective: From Simple Beginnings to Complex Systems

The history of IC packaging is a journey that mirrors the rapid evolution of electronics [1]. In the early days, around the 1960s, the first integrated circuits emerged as tiny electronic brains made of just a few transistors. These transistors worked together to perform basic calculations and control signals in electronic devices. Since the chips were relatively simple, packaging requirements were basic as well—primarily focused on protecting the fragile components from dust, moisture, and mechanical damage.

At that time, IC packaging was often a ceramic or plastic case with metal pins sticking out like legs. These pins, known as leads, connected the chip inside the package to a circuit board, allowing it to interact with other electronic components. The packaging served as a basic barrier, ensuring that the sensitive chip could function without being exposed to the harsh external environment.

As time progressed, so did the complexity of ICs. The number of transistors on a chip began to double approximately every two years, a trend known as Moore's Law [2]. By the 1980s, chips contained thousands of transistors and were used in increasingly sophisticated devices like personal computers and digital cameras. This exponential growth in chip complexity led to new demands on packaging. Not only did the package need to protect the chip, but it also had to support higher electrical performance and help manage heat generated by the chip's operation.

This period saw a shift from traditional through-hole packaging—where pins passed through holes in the circuit board—to surface-mount technology (SMT). SMT allowed components to be placed directly onto the board's surface, significantly reducing the size of electronic devices. This was a major breakthrough, as it enabled the development of smaller, more portable products like early laptops and handheld gaming consoles.

© The Author(s), under exclusive license to Springer Nature Switzerland AG 2025   1
N. Asadizanjani et al., *Introduction to Microelectronics Advanced Packaging Assurance*,
Synthesis Lectures on Engineering, Science, and Technology,
https://doi.org/10.1007/978-3-031-86102-4_1

By the 1990s and early 2000s, the push for even smaller and more powerful electronics intensified. The introduction of ball grid array (BGA) packages marked another key development. Instead of using long metal pins, BGA packages used tiny solder balls arranged in a grid pattern on the underside of the chip as mentioned in the Fig. 1.1. This innovation allowed for a higher density of connections, which was crucial for modern processors and memory chips that required thousands of electrical connections to function efficiently. BGA also helped improve the thermal and electrical performance of the packages, addressing the growing need for better heat management and faster data transmission.

The evolution didn't stop there. As the demand for miniaturization and performance continued, the industry developed new packaging techniques like chip-scale packaging (CSP)

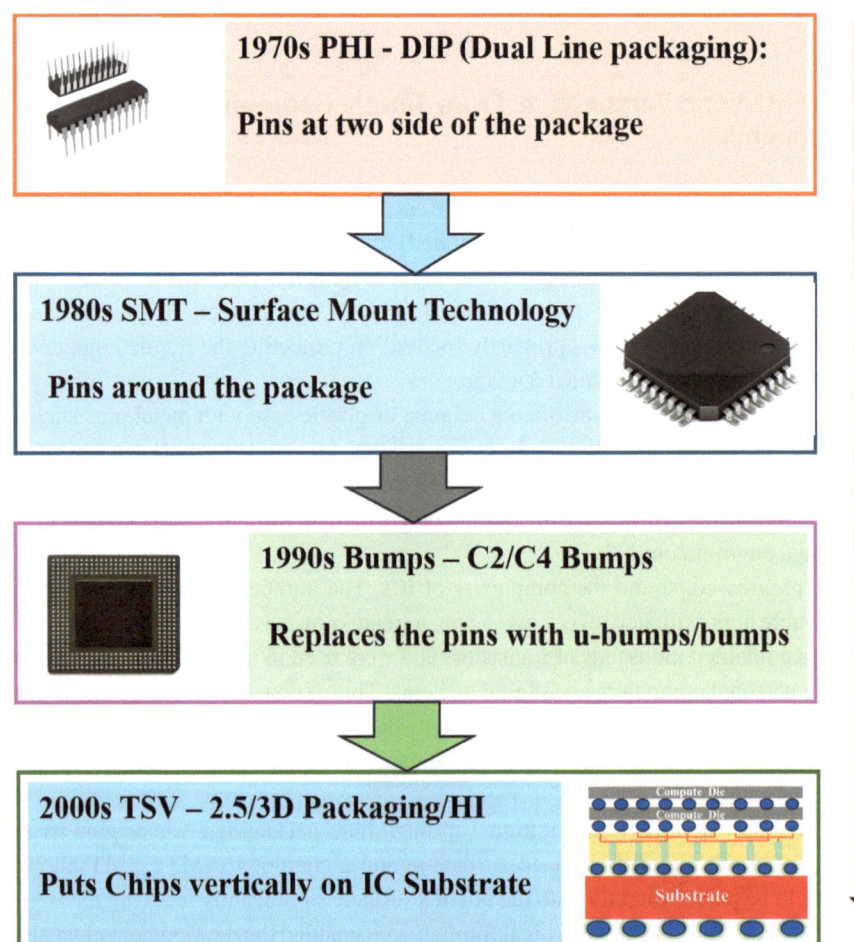

Fig. 1.1 Advancements in semiconductor packaging technology

[3], where the package size is nearly the same as the chip itself. This helped make mobile devices even smaller, without sacrificing processing power.

The trend toward more advanced packaging methods continued with 3D packaging [4], where chips are stacked vertically, like layers in a cake. By stacking multiple chips, such as processors, memory, and sensors, in a single package, the distance between them is minimized, resulting in faster communication and reduced power consumption. This approach is used in high-performance applications, including smartphones, graphics cards, and data centers, where saving space and enhancing performance are critical.

Today, packaging goes beyond just protection and connection; it integrates multiple functionalities directly within the package. Techniques like system-in-package (SiP) and heterogeneous integration [5] allow different types of chips—such as processors, memory, and communication chips—to be combined in a single package. This brings a host of benefits, from reducing the overall size of devices to improving performance and energy efficiency.

The history of IC packaging is a story of adapting to the ever-increasing demands of electronics. Each leap in technology—from through-hole pins to surface-mount technology, from ball grid arrays to 3D stacking—represents the industry's response to making devices smaller, faster, and more powerful. As we look to the future, packaging will continue to evolve, playing a key role in shaping the next generation of electronic devices.

1.2   Navigating Challenges in IC Packaging

The development of IC packaging has been a continuous journey of solving problems and overcoming challenges. As chips became smaller, faster, and more complex, packaging needed to keep up with these advancements to ensure that the chips would work reliably in various electronic devices. Each new generation of technology brought unique challenges that required innovative solutions in packaging design, materials, and manufacturing processes [6].

Managing Heat: One of the earliest and most persistent challenges in IC packaging is heat management. As chips perform more tasks and operate at higher speeds, they generate more heat. Excessive heat can damage the chip, cause it to perform poorly, or even lead to complete failure. Early solutions involved using materials that could withstand high temperatures and transfer heat away from the chip. For instance, ceramic packages were commonly used in the past because of their ability to dissipate heat.

As chips became more powerful, additional techniques such as thermal interface materials, heat sinks [7], and cooling fans were employed to keep temperatures in check. In modern packaging, advanced materials like copper heat spreaders and even liquid cooling systems are used in high-performance applications, such as gaming consoles and data center processors, to effectively manage heat.

Maintaining Electrical Performance: Another challenge was maintaining good electrical performance as chips became smaller and had more transistors packed into them. When signals travel across longer distances or encounter multiple connections, they can become distorted or delayed, which affects the chip's performance. In the early days, this wasn't a major issue, but as technology advanced, signal integrity became critical [8].

To solve this, engineers developed techniques to reduce the length of electrical pathways and increase the density of connections. One such innovation was the shift to BGA packages, where small solder balls under the chip provided numerous connection points in a compact form. This helped shorten the signal pathways, improving speed and reducing the risk of electrical interference.

Reducing Size While Increasing Functionality: As electronic devices, especially portable ones like smartphones and wearables, became more popular, the demand for smaller packages that could still perform at a high level increased. This presented a new challenge that is how to fit more functionality into a smaller space. Traditional packaging methods were no longer sufficient, as they took up too much room on the circuit board.

To address this, packaging technologies like CSP were introduced. CSP allows the package to be nearly the same size as the chip itself, drastically reducing the space it occupies on the circuit board. Additionally, advanced techniques like 3D packaging, where multiple chips are stacked on top of each other, were developed to integrate more functionality within a small footprint. This approach enables faster communication between the chips and reduces the amount of power needed, making it suitable for high-performance applications like mobile processors and artificial intelligence (AI) accelerators [9].

Improving Reliability: Packaging not only needs to make a chip work but also ensure that it keeps working over time. Factors like mechanical stress [10], temperature changes, and exposure to moisture can degrade the chip's performance or cause failures. In response to these risks, engineers developed more robust packaging materials, such as epoxy resins [11], that could better protect the chip from environmental factors. They also introduced stress-relieving structures within the package to absorb mechanical shocks and temperature-induced expansion.

Innovations in Interconnection Methods: Connecting the chip to the outside world has always been a key function of packaging. In the past, this was done using wire bonding, where fine gold or aluminum wires connected the chip to the package leads. However, as chips grew in complexity, the limitations of wire bonding—such as the time-consuming process and limited number of connections—became apparent.

Flip-chip technology [12] emerged as a solution. In flip-chip packaging, the chip is flipped upside down, and tiny solder bumps directly connect it to the package. This technique allows for a higher density of connections in a smaller space and provides better electrical performance by reducing the length of the connections. It also helps with heat dissipation, as the backside of the chip can be exposed to a heat sink [13] or other cooling methods [14].

Addressing New Challenges with Advanced Materials and Processes: With the rise of new technologies such as 5G, AI, and autonomous vehicles, packaging has to keep pace with the requirements for higher speed, lower power consumption, and greater reliability. New materials like high-density organic substrates [15] and silicon interposers [16] are now being used to support these advanced applications. Additionally, processes like wafer-level packaging [17], where the packaging steps are done directly on the wafer before it is cut into individual chips, have been developed to further miniaturize devices and improve performance.

The history of IC packaging is a story of constant innovation to overcome the challenges posed by evolving technology. From dealing with heat management to improving electrical connections and reducing size, each challenge has led to new solutions that have paved the way for the next generation of electronic devices.

1.3   The Path to Packaging: Key Steps in IC Assembly

As ICs continue to evolve, so do the processes used to assemble and package them. In modern electronics, assembly is not just about placing a chip into a protective package; it involves integrating multiple functionalities, managing heat, and ensuring that connections between different components are both efficient and reliable. The following are some of the key assembly processes used today that push the boundaries of what ICs can achieve.

System-in-Package (SiP): Combining Multiple Functions One of the major advancements in modern IC assembly is the development of SiP technology. In a traditional packaging approach, each chip—such as the processor, memory, and communication modules—would be packaged separately and then connected on a circuit board. SiP changes this by integrating multiple chips into a single package, allowing them to work together as a complete system as mentioned in Fig. 1.2.

For example, a smartphone's SiP might include the application processor, memory, power management circuits, and wireless communication modules all in one package. This integration reduces the amount of space needed inside the device, allowing for thinner and lighter products. Additionally, the shorter connections between the chips in a SiP lead to faster communication and lower power consumption, which is crucial for portable devices where battery life is a key concern.

3D Packaging: Stacking for Performance Another significant innovation in modern IC assembly is 3D packaging, where chips are stacked on top of each other, similar to the layers of a cake. This approach allows for more components to be integrated in a smaller footprint, enabling compact devices with high performance. In a 3D package, multiple chips can be connected using vertical interconnects, called through-silicon vias (TSVs), which provide a direct path for electrical signals to travel between the layers.

3D packaging is particularly useful in applications where high data transfer speeds are required, such as graphics processing units (GPUs) for gaming [18] and AI, or in memory modules for data centers. By stacking memory chips or combining memory and processors in a 3D configuration, devices can achieve higher bandwidth [19] and faster data processing while occupying less space. The Fig. 1.3 illustrates an typical example of an 3D Packaging Technology.

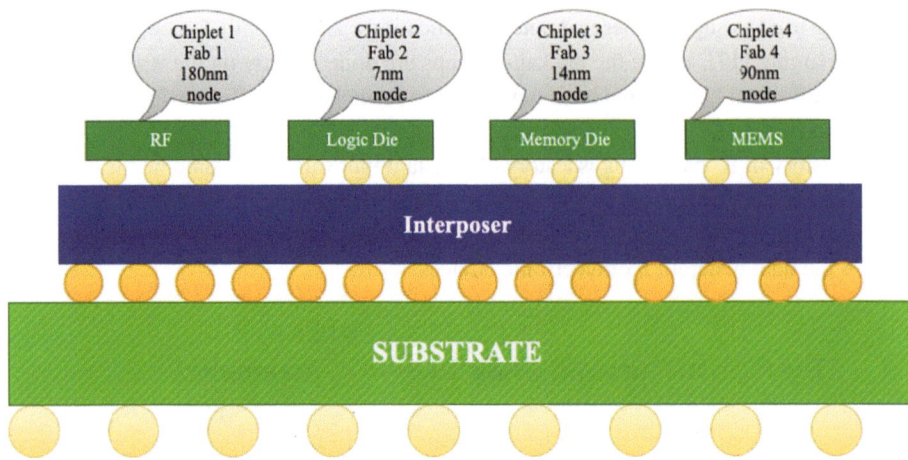

Fig. 1.2 System-in-Package (SiP) for Heterogeneous Integration (HI) (*Figure source* Noor, R., Kottur, H.R., Craig, P.J., Biswas, L.K., Khan, M.S.M., Varshney, N., Dalir, H., Akçalı, E., Motlagh, B.G., Woychik, C. and Yoon, Y.K., 2023. Us microelectronics packaging ecosystem: Challenges and opportunities. arXiv preprint arXiv:2310.11651)

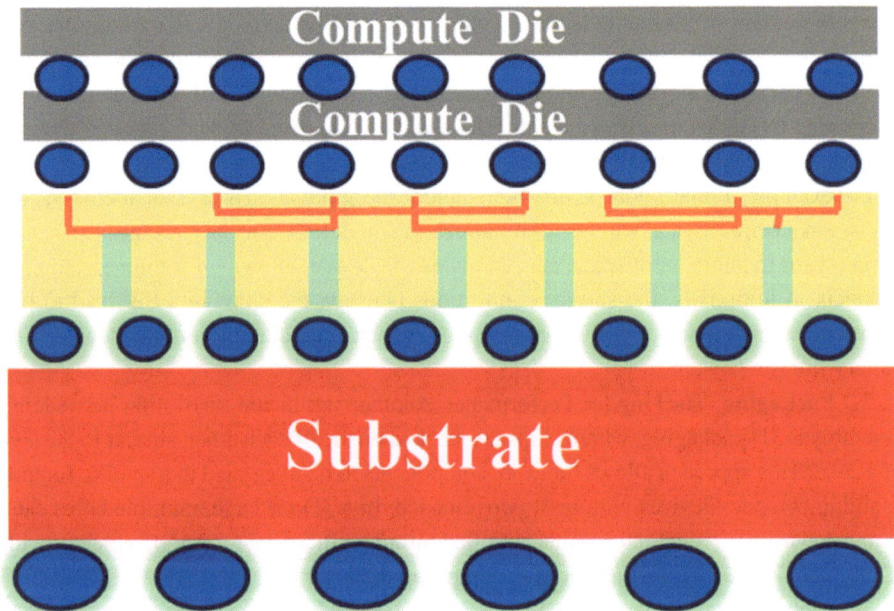

Fig. 1.3 Illustrative example of 3D packaging technology

Wafer-Level Packaging (WLP): A Step Toward Miniaturization Wafer-level packaging (WLP) is an assembly process where the packaging steps are performed while the chips are still part of the silicon wafer, before being cut into individual units. This technique enables the creation of extremely small packages, which is ideal for applications where space is at a premium, such as wearable devices, hearing aids, and tiny sensors used in the IoT.

WLP offers several advantages, including reduced package size, lower manufacturing costs, and improved electrical performance due to shorter interconnections. By packaging the chip at the wafer level, manufacturers can simplify the assembly process and achieve higher production yields, which is beneficial for mass production of small, low-power devices.

Flip-Chip Technology: Direct Connections for Efficiency Flip-chip technology has become a standard in modern IC assembly due to its ability to provide a high density of electrical connections in a compact form. In flip-chip assembly, the chip is flipped upside down so that the connections (bumps of solder) directly contact the package substrate. This approach reduces the length of electrical pathways, resulting in faster signal transmission and improved thermal management.

Flip-chip technology is widely used in high-performance applications, such as processors in computers, servers, and mobile devices, where electrical performance and heat dissipation are critical. The direct connection also allows for a larger number of inputs and outputs, which is necessary for chips with many functions and complex architectures.

Microelectromechanical Systems (MEMS) Packaging: Precision and Protection Modern IC assembly is not limited to electronic chips; it also extends to microelectromechanical systems (MEMS) [20], which are tiny mechanical devices that perform functions such as sensing, actuation, or signal processing. MEMS devices are found in products like accelerometers in smartphones, pressure sensors in cars, and microphones in hearing aids. Packaging MEMS devices presents unique challenges, as the package must protect delicate mechanical structures while still allowing them to interact with the environment.

To meet these requirements, MEMS packaging uses specialized techniques that enable the device to perform its intended function while shielding it from contaminants and mechanical damage. For example, some MEMS packages have tiny openings that allow air pressure to reach the sensor without letting dust or moisture inside. The packaging materials and assembly processes must be carefully chosen to ensure that the MEMS device remains reliable over its lifetime.

1.3.1   Steps Involved in IC Packaging

IC packaging is a multi-step process that transforms semiconductor wafers into individual chips ready for assembly in electronic products. Here's an overview of the key steps involved:

1. Wafer Dicing (Die Singulation): After the integrated circuits are fabricated on a silicon wafer, the wafer is diced into individual dies (chips). This process is known as die singulation [21]. The wafer is cut using precision tools, such as a diamond blade saw or a laser, to separate each die. Care is taken to ensure that the delicate structures on the chips are not damaged during the cutting process.

2. Die Attach: Once the dies are separated, each one is attached to a package substrate or a lead frame using a die attach [22] material, typically a conductive adhesive or solder paste. This step secures the chip in place and ensures a good thermal and electrical connection to the package, which is essential for heat dissipation and signal transmission.

3. Wire Bonding or Flip-Chip Attachment: After die attach, electrical connections need to be established between the chip and the package. This can be done using one of two methods:

a. Wire Bonding: Fine gold or aluminum wires are used to connect the bond pads on the chip to the leads of the package. Wire bonding [23] is the traditional and most widely used method for making electrical connections.

b. Flip-Chip Attachment: In this process, the chip is flipped upside down, and tiny solder bumps on the chip surface are aligned with corresponding pads on the package substrate. The solder bumps are then reflowed to create direct electrical connections. Flip-chip technology is used for applications that require high connection density and better thermal management.

4. Encapsulation (Molding): The chip and the bond wires (if present) are encapsulated in a protective material, usually an epoxy resin, to shield them from physical damage, moisture, and other environmental factors. The encapsulation process [24], also known as molding, involves placing the assembly in a mold cavity and injecting the encapsulant material around the chip. For some advanced packages, such as FOWLP, a different type of encapsulation process is used to embed the chip within the mold compound.

5. Plating: After encapsulation, the external leads or contact areas of the package are plated with a thin layer of metal, typically a combination of nickel, palladium, and gold, or other materials. Plating improves the solderability of the leads, ensuring good electrical connections when the chip is mounted onto a circuit board [25].

6. Singulation (Package Separation): For multi-chip packages or when packages are processed on a panel or wafer level, the individual packages need to be separated. This step, similar to wafer dicing, involves cutting the panel or wafer to create discrete package units. The process is done using saws or lasers, depending on the type of package.

7. Marking: Each package is marked with information such as the part number, manufacturer's logo, and date code. Marking is usually done using laser etching or ink printing. This step ensures traceability and identification of the chip for quality control and inventory management [26].

8. Testing: Before final assembly, the packaged ICs undergo rigorous testing to check for functionality and quality. Testing includes verifying electrical performance, checking for defects, and ensuring the package meets the required specifications [27]. This is typically done using automated test equipment (ATE), which can test multiple chips simultaneously [28].

9. Taping and Reeling: Once the packages have passed the testing phase, they are placed into protective carrier tapes. These tapes are then wound onto reels for easy handling, trans-

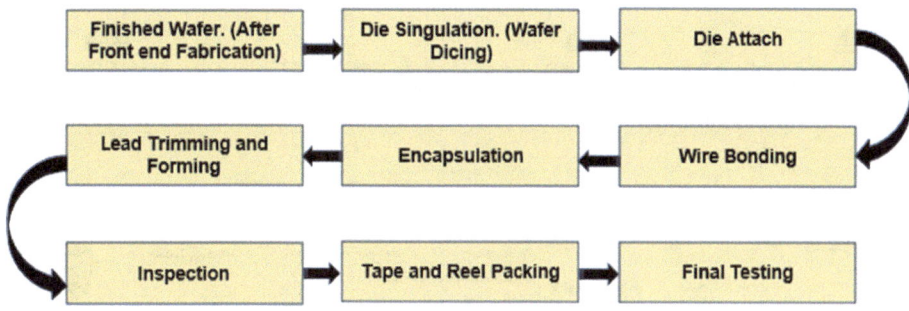

Fig. 1.4 Key steps in the semiconductor packaging assembly process

portation, and automatic assembly in downstream manufacturing processes. Taping and reeling help protect the chips from damage during storage and facilitate high-speed automated assembly in printed circuit board (PCB) production [29]. Keysteps of the packaging assembly is illustrated below in Fig. 1.4.

1.4   The Complex Ecosystem of Semiconductor Packaging

The semiconductor packaging ecosystem as depicted in the Fig. 1.5 is a multifaceted and global network involving multiple stages, from material sourcing to final product assembly. It encompasses a wide range of processes, technologies, and stakeholders, including raw material suppliers, equipment manufacturers, outsourced semiconductor assembly and test (OSAT) providers, and integrated device manufacturers (IDMs). This ecosystem is not only about assembling chips into functional units but also ensuring electrical, mechanical, and thermal performance while maintaining high levels of reliability and security [30].

The semiconductor packaging industry is a multi-billion dollar market that continues to grow due to increasing demand for advanced electronics, autonomous systems, and artificial intelligence. In 2023, the global semiconductor packaging market was valued at approximately $69 billion and is projected to reach $85 billion by 2028, growing at a compound annual growth rate (CAGR) of 4.2%. This growth is driven by emerging packaging technologies like SiP, 2.5D/3D packaging, and advanced packaging techniques required for heterogeneous integration (HI).

Packaging techniques range from traditional wire-bonding to flip-chip and wafer-level packaging and its ecosystem is mentioned in Fig. 1.6. The choice of technology is often determined by the application's requirements for performance, cost, and size.

Advanced packaging now accounts for more than 50% of the packaging market due to the demand for high-performance computing and communication devices, including smartphones, servers, and automotive applications.

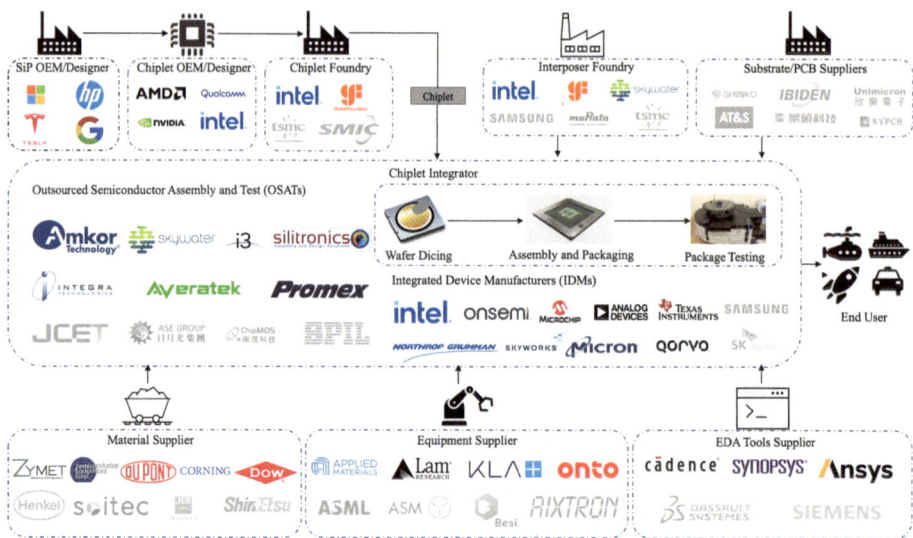

Fig. 1.5 US-Based Advanced Packaging Supply Chain with Onshore and Offshore Companies (Offshore Indicated in Grey). (*Figure source* Noor, R., Kottur, H.R., Craig, P.J., Biswas, L.K., Khan, M.S.M., Varshney, N., Dalir, H., Akçalı, E., Motlagh, B.G., Woychik, C. and Yoon, Y.K., 2023. Us microelectronics packaging ecosystem: Challenges and opportunities. arXiv preprint arXiv:2310.11651)

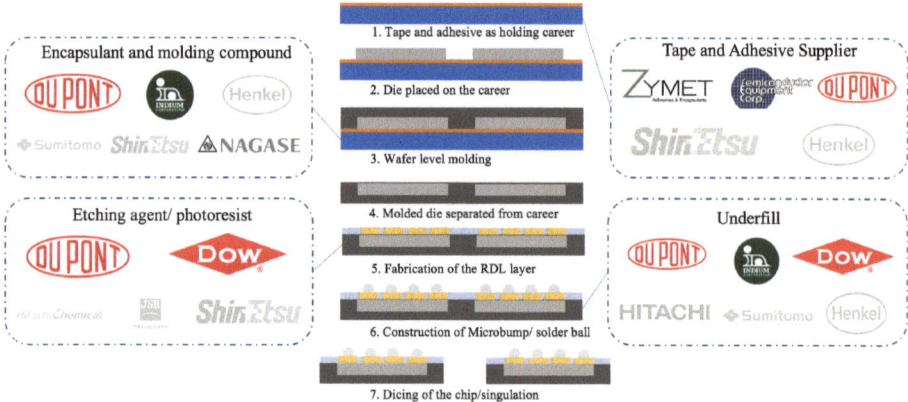

Fig. 1.6 Typical wafer-level packaging process flow with onshore and offshore (greyed out logo) material suppliers. (*Figure source* Noor, R., Kottur, H.R., Craig, P.J., Biswas, L.K., Khan, M.S.M., Varshney, N., Dalir, H., Akçalı, E., Motlagh, B.G., Woychik, C. and Yoon, Y.K., 2023. Us microelectronics packaging ecosystem: Challenges and opportunities. arXiv preprint arXiv:2310.11651)

Layers of Complexity in Packaging The packaging process involves multiple layers, each adding a degree of complexity:

Materials Supply Chain: The packaging process relies on a variety of materials, including silicon wafers, bonding wires, organic substrates, lead frames, and advanced materials like low-k dielectrics and thermal interface materials. Any disruption in the supply of these materials can affect the entire packaging process [31].

Equipment Manufacturers: The machinery used for processes like die attach, wire bonding, flip-chip placement, and wafer dicing is highly specialized. The equipment industry, worth approximately $8 billion in 2023, is dominated by a few key players, making it vulnerable to supply chain disruptions.

Outsourced Semiconductor Assembly and Test (OSAT): The OSAT market was valued at $33 billion in 2023. OSAT companies provide back-end services like assembly, packaging, and testing, handling around 70% of global packaging. These companies often operate offshore facilities in regions like Taiwan, China, and Southeast Asia, introducing geopolitical and logistical challenges [32].

Technological Innovation: The need for advanced packaging techniques to support heterogeneous integration and high-performance computing adds complexity. Technologies like 2.5D/3D packaging and chiplets integration require sophisticated interposer designs, new materials, and enhanced testing protocols.

1.4.1   The US Semiconductor Packaging Ecosystem: Challenges and Specific Metrics

Overview of the US Market The U.S. holds a strong position in semiconductor design and manufacturing, with about 47% of global market share in semiconductor sales as of 2023. However, the domestic semiconductor packaging and assembly capabilities are relatively underdeveloped compared to global competitors, particularly in advanced packaging. The U.S. accounts for only about 12% of the world's advanced packaging capacity, lagging behind Asian countries like Taiwan (42%) and South Korea (18%). Advanced packaging market share is depicted in Fig. 1.7 as of 2020.

Key Challenges in the US Packaging Ecosystem Limited Domestic Advanced Packaging Infrastructure:
Despite the U.S. being a leader in semiconductor design and front-end manufacturing, it significantly trails in back-end processes, including advanced packaging and OSAT services. For instance, over 80% of OSAT services utilized by U.S.-based semiconductor companies are sourced from overseas, mainly from Taiwan and China. This heavy reliance on offshore facilities introduces vulnerabilities related to supply chain security and potential geopolitical risks. Efforts to bring more packaging capabilities onshore are ongoing, with government initiatives like the CHIPS Act, which allocates $52 billion to revitalizing the U.S. semiconductor industry, including investments in advanced packaging. However, scaling domestic infrastructure to match overseas capabilities remains a considerable challenge.

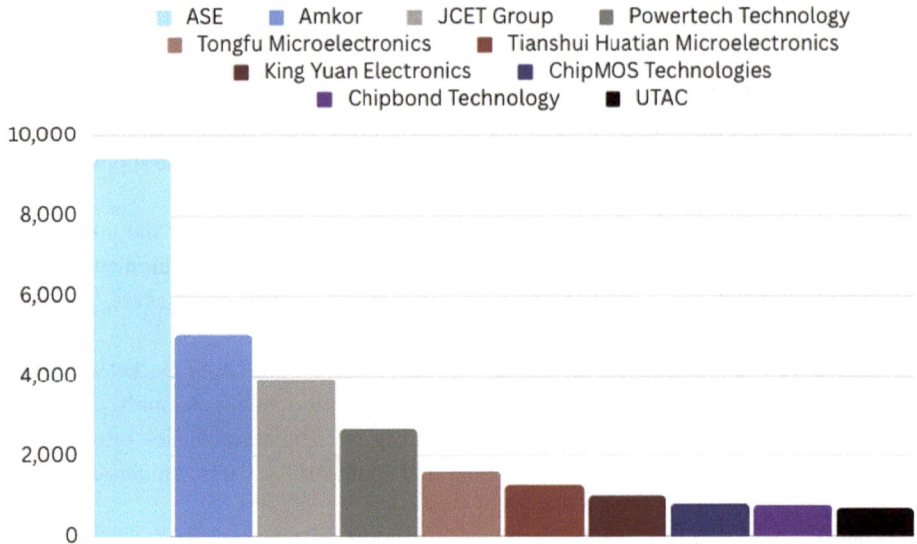

Fig. 1.7 2020 advanced packaging market share in US$ million. (*Figure source* Noor, R., Kottur, H.R., Craig, P.J., Biswas, L.K., Khan, M.S.M., Varshney, N., Dalir, H., Akçalı, E., Motlagh, B.G., Woychik, C. and Yoon, Y.K., 2023. Us microelectronics packaging ecosystem: Challenges and opportunities. arXiv preprint arXiv:2310.11651)

High Dependence on Offshore OSAT Facilities:

The U.S. packaging industry heavily depends on offshore OSAT providers, which hold approximately 70% of the global OSAT market share. The concentration of OSAT facilities in Asia makes the supply chain vulnerable to disruptions caused by geopolitical tensions and natural events or other regional instabilities. Given that many of these facilities operate in countries with different regulatory frameworks and standards for hardware security, concerns around the potential for tampering, reverse engineering, or the introduction of counterfeit components are prevalent [33].

Technological Bottlenecks:

The U.S. faces significant technological challenges in scaling up its advanced packaging capabilities. While traditional packaging methods (e.g., wire-bonding, plastic molding) are well established domestically, advanced techniques such as 2.5D/3D integration, TSV, and hybrid bonding are less prevalent in U.S. facilities. For example, Taiwan and South Korea have invested heavily in 2.5D/3D packaging technologies, with companies like TSMC and Samsung leading the field. The U.S., on the other hand, lacks equivalent large-scale advanced packaging facilities. The development of advanced interconnect technologies and materials for high-density integration is another bottleneck. Materials like silicon interposers, copper pillars, and advanced thermal interface materials are critical for packaging technologies supporting high-performance computing.

Security and Assurance in the Supply Chain:
A critical aspect of the U.S. packaging ecosystem is ensuring hardware security and trust. With over 75% of global advanced packaging performed in offshore facilities, it becomes challenging to maintain control over the integrity of the packaged products. Advanced packaging processes, which involve numerous stages and facilities, can be exploited for inserting hardware Trojans or counterfeit components. Establishing a trusted supply chain requires enhancing the traceability of materials and components across all stages of packaging.

1.4.2 Domestic Expansion of Advanced Packaging Capabilities

The U.S. has opportunities to expand its domestic advanced packaging infrastructure. With federal investments, such as the CHIPS Act, aimed at boosting semiconductor manufacturing and packaging capabilities, there is a push towards building new advanced packaging facilities and retrofitting existing ones [34]. Companies like Intel and SkyWater Technology are already taking steps in this direction. For instance, Intel has announced plans to develop advanced packaging capabilities at its facilities in Oregon and Arizona, focusing on technologies like EMIB (Embedded Multi-die Interconnect Bridge) and Foveros [35].

Enhanced Research and Development:
The U.S. can strengthen its position in advanced packaging by investing in R&D for novel materials, interconnect technologies, and packaging methods. In 2023, the U.S. government allocated over $1 billion for semiconductor packaging research, aiming to overcome current technological limitations and develop next-generation packaging solutions. Collaborative efforts between academia, industry, and government play a pivotal role in fostering innovations to tackle key challenges in the semiconductor packaging domain. These include addressing heat dissipation in high-performance packages, establishing standardized testing protocols for multi-chip modules, and enhancing yield rates for complex packaging designs.

Workforce Development Initiatives:
Workforce development is vital to reinforcing the semiconductor packaging ecosystem, especially in the United States, where the demand for skilled professionals continues to grow alongside advancements in packaging technologies. Building a competent and well-equipped workforce is key to ensuring the industry keeps up with rapid technological progress while maintaining a strong global competitive position. Below is an in-depth exploration of initiatives and strategies aimed at advancing workforce development in semiconductor packaging:

    1. Collaboration Between Industry and Academia
To address the skills gap, collaboration between the semiconductor industry and educational institutions is essential:

    University Programs and Curricula Development: Colleges and universities need to create or expand programs in microelectronics, packaging technology, materials science, and electrical engineering with a specific focus on semiconductor packaging. This includes updating curricula to reflect the latest industry trends, such as heterogeneous integration (HI), 2.5D/3D packaging, and advanced interconnect technologies.

Industry-Sponsored Research and Development: Companies can partner with academic institutions to fund research programs that tackle real-world challenges in semiconductor packaging. This provides students with hands-on experience and exposure to cutting-edge technologies, while companies benefit from fresh ideas and potential future hires.

Internship and Co-op Programs: Establishing robust internship and cooperative education programs allows students to gain practical experience in semiconductor packaging facilities. These programs help bridge the gap between theoretical knowledge and practical application, making students more job-ready upon graduation.

2. Government-Funded Workforce Development Programs

Government initiatives can significantly boost workforce development by providing funding and policy support for education and training programs:

CHIPS and Science Act Initiatives: In the U.S., the CHIPS and Science Act includes provisions to boost semiconductor manufacturing and R&D, with a portion dedicated to workforce development and a few other initiatives as depicted in Fig. 1.8. It encourages the establishment of regional hubs for semiconductor education, providing funding for training programs, scholarships, and apprenticeships.

Grants for Technical Training Centers: Government grants can support the establishment of specialized technical training centers that focus on semiconductor packaging skills. These centers can offer short-term certification programs, skills training for technicians, and continuing education for engineers in the workforce.

Apprenticeship Programs: Formal apprenticeship programs, supported by government funding, can provide a structured pathway for students and young professionals to gain

Initiative/Policy	Key Focus Areas	Year	Description
CHIPS and Science Act of 2022	Manufacturing, Workforce Development	2022	Allocates $52.7 billion for semiconductor research, development, and manufacturing. Includes incentives for building new facilities and advancing packaging technologies.
National Semiconductor Technology Center (NSTC)	Research, Innovation, Collaboration	2022	Established under the CHIPS Act to drive semiconductor research, innovation, and collaboration between industry, academia, and government.
National Advanced Packaging Manufacturing Program (NAPMP)	Advanced Packaging, Manufacturing	2022	Focuses on developing advanced packaging technologies, enhancing U.S. capabilities in semiconductor packaging.
Trusted Foundry Program	National Security, Domestic Manufacturing & Ongoing	Ongoing	Ensures a secure and reliable supply of semiconductors for defense applications through domestic foundries.
Microelectronics Commons Initiative	Defense, Semiconductor Technology Development	Ongoing	Accelerates the development of critical semiconductor technologies essential for national defense.
America's Foundries Act	Domestic Manufacturing, Infrastructure	2020	Provides funding and incentives for building new semiconductor foundries and expanding domestic manufacturing capabilities.
U.S. Innovation and Competition Act (USICA)	R&D, Innovation, Workforce Development	2021	Supports semiconductor R&D, innovation, and workforce development to maintain U.S. competitiveness globally.

Fig. 1.8 Summary of U.S. Government initiatives and policies for advanced packaging and semiconductors

hands-on experience in semiconductor packaging. Apprenticeships combine classroom instruction with paid on-the-job training, making them accessible for individuals who may not pursue traditional college degrees.

3. On-the-Job Training and Upskilling Programs For the existing workforce, continuous training and skill development are vital to keeping up with evolving technologies:

Corporate Upskilling Initiatives: Companies can develop internal training programs to upskill their current workforce in new packaging techniques, such as advanced lithography, hybrid bonding, and AI-driven packaging solutions. Providing employees with opportunities for ongoing learning can increase retention and improve productivity.

Cross-Training Programs: Implementing cross-training initiatives within companies allows employees to develop multiple skill sets, such as transitioning from process engineering to quality assurance in packaging. This not only enhances employee versatility but also helps companies adapt to changing demands.

Use of Online Learning Platforms: Online courses and digital platforms can complement traditional training by providing flexible learning opportunities. Platforms like Coursera, edX, and company-specific learning management systems (LMS) can offer courses on semiconductor fundamentals, packaging design, materials science, and emerging technologies.

4. Outreach Programs to K-12 Students

Building interest in semiconductor packaging at an early age is crucial to developing a sustainable pipeline of future talent:

STEM Outreach Programs: Initiatives such as science fairs, engineering camps, and robotics competitions can help inspire K-12 students to pursue careers in microelectronics and packaging. Companies and universities can sponsor these events or provide guest speakers to talk about semiconductor technologies.

Curriculum Integration: Incorporating semiconductor-related content into K-12 curricula can spark interest early on. Programs can include hands-on activities like basic circuit assembly, microchip design workshops, and introductory lessons on how everyday electronic devices are made.

Dual-Enrollment Programs: High school students can participate in dual-enrollment programs that allow them to take college-level courses in electronics, materials science, or mechanical engineering. This gives them a head start on learning the skills required for careers in semiconductor packaging.

5. Increasing Diversity in the Workforce

Ensuring a diverse workforce in semiconductor packaging not only addresses talent shortages but also brings different perspectives and ideas to the industry:

Targeted Scholarships and Mentorship Programs: Scholarships for underrepresented groups (women, minorities, veterans) in STEM fields can make education in microelectronics and packaging more accessible. Mentorship programs can provide support and guidance for these students, increasing their retention in the field.

Inclusive Hiring Practices: Companies can implement policies to increase diversity in hiring for engineering roles, technicians, and management positions within the packaging industry. This may involve partnerships with organizations that focus on diversity in STEM, such as the Society of Women Engineers (SWE) or the National Society of Black Engineers (NSBE).

Employee Resource Groups (ERGs): Establishing ERGs within companies can foster a sense of community among diverse employees and provide a platform for professional development, networking, and advocating for diversity initiatives.

6. Certifications and Industry Standards

Establishing recognized certifications and adhering to industry standards can validate skills and ensure consistent workforce quality:

Industry Certification Programs: Certifications such as IPC (Institute for Interconnecting and Packaging Electronic Circuits) or SEMI (Semiconductor Equipment and Materials International) standards can provide recognition for specialized skills in semiconductor packaging. Companies can encourage employees to obtain these certifications to ensure a baseline level of competence.

Standardized Training Programs: Developing industry-wide training programs that meet specific standards can help ensure that employees across the sector have consistent skills and knowledge. This is especially important for areas like cleanroom protocols, equipment maintenance, and quality assurance in packaging processes.

7. Regional Workforce Development Hubs Establishing regional hubs focused on semiconductor education and workforce training can bring together educational institutions, industry players, and government agencies:

Public-Private Partnerships: Creating workforce development hubs through public-private partnerships can facilitate the sharing of resources, infrastructure, and expertise. For instance, these hubs can provide training equipment, cleanroom facilities, and hands-on labs that simulate real-world semiconductor packaging environments.

Specialization by Region: Different regions can specialize in various aspects of the semiconductor industry, such as chip design, packaging, testing, or manufacturing. For example, one region may focus on advanced packaging techniques, while another specializes in semiconductor materials research. This approach can create a comprehensive national network of semiconductor expertise.

Strengthening Hardware Security Protocols:

Securing the semiconductor packaging lifecycle is critical to maintaining the integrity and reliability of electronic components, particularly as supply chain complexity and security threats continue to grow. One approach involves developing tamper-evident packaging techniques that make unauthorized access detectable. This can be achieved using physical indicators, such as seals that change color when disturbed, embedded sensors that detect changes in temperature or electromagnetic fields, or tamper-resistant materials that obscure internal circuitry. These measures not only deter tampering but also provide clear evidence of intrusion attempts. Additionally, secure hardware designs can enhance the overall protection by integrating cryptographic modules, secure enclaves, and anti-counterfeiting features like Physically Unclonable Functions (PUFs) directly within the semiconductor package. These features create a robust line of defense against tampering, cloning, and reverse engineering [36].

Monitoring supply chain integrity involves implementing comprehensive protocols to track components from fabrication to final assembly. Blockchain technology can be a game-changer in this area by enabling decentralized, tamper-proof records of each step in the supply chain. Using a blockchain ledger ensures that any attempts to alter records would be easily detectable, increasing transparency and trust among stakeholders. Meanwhile, AI-based security screening can analyze data from inspection processes, such as X-ray images or electrical tests, to identify counterfeit or substandard components [37]. Machine learning algorithms can detect subtle inconsistencies or patterns that indicate potential threats, thereby enhancing the accuracy and efficiency of counterfeit detection. Together, these technologies can significantly bolster the security of the semiconductor packaging lifecycle, addressing risks from tampering and supply chain vulnerabilities.

Conclusion:

In this chapter, we explored the evolution of IC packaging, from traditional methods to advanced techniques, highlighting the challenges and innovations shaping the field. We examined the various packaging processes, including die singulation, bonding, and final assembly, while emphasizing the significance of a robust packaging ecosystem, particularly in the U.S. Additionally, we discussed workforce development initiatives aimed at bridging the skills gap and fostering talent in semiconductor packaging. The importance of securing the packaging lifecycle through tamper-evident measures, blockchain traceability, and AI-based counterfeit detection was also emphasized. Together, these insights underscore the critical role of advanced packaging in driving semiconductor progress and ensuring a secure, resilient supply chain.

References

1. Jin, Haocheng. (2023). The History, Current Applications and Future of Integrated Circuit. *Highlights in Science, Engineering and Technology*, **31**, 232–238. https://doi.org/10.54097/hset.v31i.5146
2. Schaller, R. R. (1997). Moore's law: past, present and future. *IEEE Spectrum*, **34**(6), 52–59. https://doi.org/10.1109/6.591665
3. Greig, W. (2007). *The Chip Scale Package*. https://doi.org/10.1007/0-387-33913-2_4
4. Bolanos, M. A. (2010). 3D Packaging Technology: Enabling the next wave of applications. In *2010 IEEE/CPMT International Electronic Manufacturing Technology Symposium (IEMT)*, Melaka, Malaysia, pp. 1–5. https://doi.org/10.1109/IEMT.2010.5746735

5. Do, W. (2018). High-Density Fan-Out Technology for Advanced SiP and Heterogeneous Integration. In *2018 IEEE 2nd Electron Devices Technology and Manufacturing Conference (EDTM)*, Kobe, Japan, pp. 138–141. https://doi.org/10.1109/EDTM.2018.8421496
6. Chen, Z., et al. (2023). Challenges and prospects for advanced packaging. *Fundamental Research*, n. pag.
7. Chowdhury, A., et al. (2024). Thermal Management and Integrated Heat Spreader Assembly Challenges of Products with Variable Die Heights. In *2024 IEEE Intersociety Conference on Thermal and Thermomechanical Phenomena in Electronic Systems (ITherm)*. IEEE.
8. Shi, H., et al. (2024). Investigation of Heat Dissipation and Electrical Properties of Diamond Interposer for 2.5-D Packagings. *IEEE Transactions on Components, Packaging and Manufacturing Technology*.
9. Kim, H. J., & Jung, J. P. (2023). Artificial Intelligence Semiconductor and Packaging Technology Trend. *Journal of the Microelectronics and Packaging Society*, **30**(3), 11–19.
10. Gan, C. L., et al. (2023). Evolution of epoxy molding compounds and future carbon materials for thermal and mechanical stress management in memory device packaging: a critical review. *Journal of Materials Science: Materials in Electronics*, **34**(30), 2011.
11. Jiao, D., et al. (2024). Engineering flame retardant epoxy resins with strengthened mechanical property by using reactive catechol functionalized DOPO compounds. *Chemical Engineering Journal*, **485**, 149910.
12. Sun, Y., et al. (2024). Flip-chip solder bumps defect detection using a self-search lightweight framework. *Advanced Engineering Informatics*, **60**, 102395.
13. Marseglia, G., et al. (2024). Enhancement of microchannel heat sink heat transfer: Comparison between different heat transfer enhancement strategies. *Experimental Thermal and Fluid Science*, **150**, 111052.
14. Li, Y., et al. (2023). Design and optimization of heat sinks for the liquid cooling of electronics with multiple heat sources: a literature review. *Energies*, **16**(22), 7468.
15. Lau, J. H., et al. (2023). Hybrid Substrate With Ultralarge Organic Interposer for Heterogeneous Integration. *IEEE Transactions on Components, Packaging and Manufacturing Technology*, **13**(9), 1371–1379.
16. Mizutani, M., et al. (2024). Study for realization of the next generation high density RDL packaging for 2.5 D large silicon interposer. In *2024 IEEE 74th Electronic Components and Technology Conference (ECTC)*. IEEE.
17. Aleksov, A., et al. (2024). Organic Interposers Using Zero-Misalignment-Via Technology and Silicon Wafer Carriers for Large Area Wafer-Level Package Applications. In *2024 IEEE 74th Electronic Components and Technology Conference (ECTC)*. IEEE.
18. Vaithianathan, M., et al. (2023). Comparative Study of FPGA and GPU for High-Performance Computing and AI. *ESP International Journal of Advancements in Computational Technology (ESP-IJACT)*, **1**(1), 37–46.
19. Novick, A., et al. (2023). High-bandwidth density silicon photonic resonators for energy-efficient optical interconnects. *Applied Physics Reviews*, **10**(4).
20. Li, R., et al. (2024). Enhancement of cooling performance in MEMS by modifying 3D packaging structure: A design, integration, analysis and test study. *Applied Thermal Engineering*, **237**, 121758.
21. Hu, Z., et al. (2024). Utilizing Generative Adversarial Networks for Image Data Augmentation and Classification of Semiconductor Wafer Dicing Induced Defects. In *2024 IEEE 29th International Conference on Emerging Technologies and Factory Automation (ETFA)*. IEEE.
22. Hou, F., et al. (2024). Review of Die-Attach Materials for SiC HighTemperature Packaging. *IEEE Transactions on Power Electronics*.
23. Zhou, H., et al. (2023). Research progress on bonding wire for microelectronic packaging. *Micromachines*, **14**(2), 432.

24. Zhou, M., et al. (2024). Co-encapsulation of anthocyanin and cinnamaldehyde in nanoparticle-filled carrageenan films: Fabrication, characterization, and active packaging applications. *Food Hydrocolloids*, **149**, 109609.

25. Chen, K.-X., et al. (2023). Research progress of electroplated nanotwinned copper in microelectronic packaging. *Materials*, **16**(13), 4614.

26. Hoveida, P., et al. (2023). Terahertz-readable laser engraved marks as a novel solution for product traceability. *Scientific Reports*, **13**(1), 12474.

27. Wang, H., et al. (2023). A review of system-in-package technologies: application and reliability of advanced packaging. *Micromachines*, **14**(6), 1149.

28. Khan, M. S. M., et al. (2023). Exploring advanced packaging technologies for reverse engineering a system-in-package (SiP). *IEEE Transactions on Components, Packaging and Manufacturing Technology*.

29. Tang, H. S.-W., et al. (2024). Packaging Challenges and Solutions for Next Generation Low-Profile WLCSP. In *2024 IEEE 74th Electronic Components and Technology Conference (ECTC)*. IEEE.

30. Noor, R., et al. (2023). US microelectronics packaging ecosystem: Challenges and opportunities. *arXiv preprint* arXiv:2310.11651.

31. Xiong, W., Wu, D. D., & Yeung, J. H. Y. (2024). Semiconductor supply chain resilience and disruption: Insights, mitigation, and future directions. *International Journal of Production Research*, 1–24.

32. Collier, Z. A., et al. (2023). Stress Testing for Resilience of Semiconductor Supply Chains. In *2023 IEEE 14th Annual Ubiquitous Computing, Electronics & Mobile Communication Conference (UEMCON)*. IEEE.

33. Haramboure, A., et al. (2023). Vulnerabilities in the semiconductor supply chain. Unpublished report.

34. Peters, M. A. (2023). Semiconductors, geopolitics and technological rivalry: the US CHIPS & Science Act, 2022. *Educational Philosophy and Theory*, **55**(14), 1642–1646.

35. Prasad, C., et al. (2020). Silicon reliability characterization of Intel's Foveros 3D integration technology for logic-on-logic die stacking. In *2020 IEEE International Reliability Physics Symposium (IRPS)*. IEEE.

36. Khan, M. S. M., et al. (2023). Exploring advanced packaging technologies for reverse engineering a system-in-package (SiP). *IEEE Transactions on Components, Packaging and Manufacturing Technology*.

37. Varshney, N., et al. (2024). Challenges and opportunities in non-destructive characterization of stacked IC packaging: Insights from SAM and 3D X-ray analysis. In *Developments in X-Ray Tomography XV*, **13152**, 92–99.

Bonding Techniques–Interconnects

2

2.1 Role of Bonding in Packaging

Historically, the development of bonding techniques in semiconductor packaging has focused on finding the optimal balance between cost, performance, and reliability. As technology has advanced, bonding methods have evolved significantly, moving from traditional approaches like wire bonding to more sophisticated techniques such as hybrid bonding and wafer-level bonding and could be visualized in Fig. 2.1. This progression has been driven by the increasing demand for smaller, faster, and more efficient electronic devices. As electronic components continue to shrink in size, bonding techniques must support fine-pitch connections to accommodate the dense integration of circuits and components. High-performance applications also place greater demands on bonding technologies, requiring low-inductance and low-resistance interconnections to ensure that signals travel quickly and with minimal loss. Additionally, trends like system-on-chip (SoC) and SiP have accelerated the shift toward advanced packaging solutions, including 2.5D and 3D configurations, which integrate multiple chips or layers into a single package to enhance functionality. These evolving requirements necessitate a variety of bonding techniques, each with specific advantages and limitations, allowing manufacturers to tailor the choice of bonding method based on the needs of different applications, such as miniaturization, performance optimization, or high-density integration.

2.2 Wire Bonding: A Longstanding and Evolving Technology (1950-Present)

Wire bonding, first developed in the 1950s, remains a foundational technique in semiconductor packaging, providing a reliable method for connecting the silicon die, which contains the integrated circuit, to a PCB or package leads [1]. This technology involves using thin

© The Author(s), under exclusive license to Springer Nature Switzerland AG 2025

N. Asadizanjani et al., *Introduction to Microelectronics Advanced Packaging Assurance*, Synthesis Lectures on Engineering, Science, and Technology, https://doi.org/10.1007/978-3-031-86102-4_2

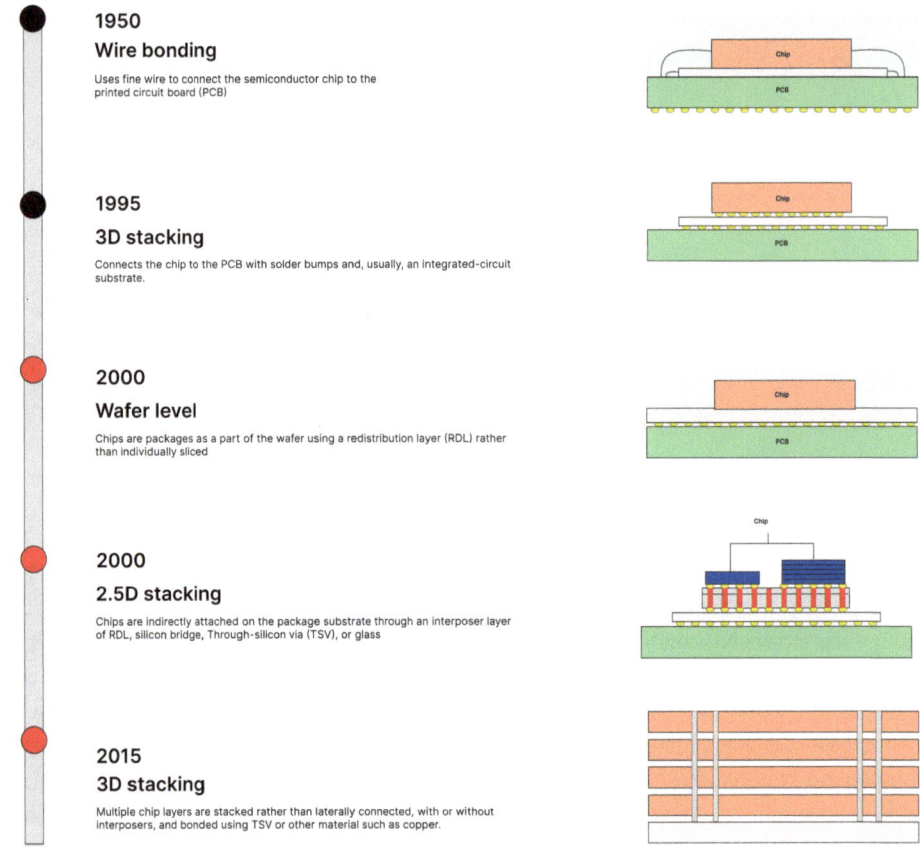

Fig. 2.1 Advancements in bonding interconnects for semiconductor packaging

metal wires, typically made of gold, aluminum, or copper, and small solder balls to form electrical connections [2]. The introduction of wire bonding revolutionized semiconductor packaging by enabling mass production of semiconductor devices with consistent quality and reliability [3]. Initially, the process was manual and labor-intensive, involving specialized tools to form individual connections [4]. However, the emergence of automated wire bonding machines in the 1960s significantly increased productivity and accuracy, making it possible to meet the growing demands of the semiconductor industry.

Over the decades, advancements in wire bonding materials [5] and techniques have improved the overall performance and reliability of this method. The transition from gold to aluminum, and later to copper wires, brought benefits such as better electrical conductivity, enhanced reliability, and reduced cost. The development of ultrasonic bonding [6], where high-frequency vibrations are used to form the bond, further strengthened the connections and made the process more robust, especially under harsh environmental conditions.

Despite these advancements, certain challenges remain, such as susceptibility to failure under high humidity [7], temperature cycling, and high-temperature conditions. These factors can impact the long-term reliability of wire-bonded packages, particularly in demanding applications.

Wire bonding continues to be widely used today, accounting for roughly 75 to 80% of all semiconductor packages, thanks to its cost-effectiveness and proven reliability. It remains the go-to solution for low-cost legacy packages, mid-range packages, and certain memory applications [8], where more complex or expensive packaging techniques may not be justified. Wire bonding is also making inroads into new applications, such as automotive electronics and sensor technology, where cost constraints and proven durability make it an attractive option. Although some traditional wire-bond applications are migrating to advanced packaging types, such as flip-chip and wafer-level packaging, wire bonding continues to grow, particularly in the area of copper wire bonding, which offers improved performance at a lower cost.

However, wire bonding does face limitations, especially when it comes to high-performance chips that require greater I/O density and bandwidth. The sequential nature of forming each bond adds complexity to the manufacturing process, potentially leading to production delays. Additionally, for advanced semiconductor nodes, where fine-pitch connections are necessary, wire bonding may not be able to meet the demands for high-frequency and high-speed applications. Nevertheless, for certain silicon nodes that cannot easily adopt advanced packaging techniques, wire bonding still provides a competitive edge in terms of cost and dependability. Thus, while its growth prospects may be tempered by the rise of newer technologies, wire bonding remains an essential and evolving part of the semiconductor packaging landscape, continuing to serve a diverse range of applications where it excels in balancing cost, performance, and reliability.

Wire bonding can be further categorized into.

2.2.1 Ball Bonding

Ball bonding is the most widely used form of wire bonding, especially for gold and copper wires [9]. The process begins with the formation of a ball at the end of the wire through electric flame-off (EFO), a step that briefly melts the tip of the wire into a spherical shape [10]. This ball is then pressed onto the bond pad of the chip or substrate using a combination of heat, pressure, and ultrasonic energy. Once the first bond is established, the wire is extended to the second bond pad, where it is attached with a wedge-like stitch bond [11]. This method allows for rapid connections, making it well-suited for high-volume production environments.

Ball bonding is widely employed in memory modules, consumer electronics, and automotive devices. It offers several advantages, such as high production speeds and the ability to create strong, reliable bonds. Gold wire, although expensive, provides excellent resistance to corrosion, while copper offers a more economical alternative, though it requires additional steps to prevent oxidation. However, ball bonding is not without limitations. It introduces longer interconnect paths, which can degrade signal quality at high frequencies, making it less suitable for GHz-range signals found in RF and 5G devices. Furthermore, as chips become smaller and more densely packed, fine-pitch bonding becomes challenging, necessitating precise alignment to prevent bridging or bond failure.

2.2.2   Wedge Bonding

Wedge bonding, on the other hand, uses a wedge-shaped tool to press the wire directly onto the bond pad without forming a ball [12]. This method is typically employed with aluminum wire, which is more affordable than gold and well-suited for low-temperature bonding processes. Wedge bonding creates interconnects by bonding the wire at one point, stretching it across the desired path, and then forming a second bond with the same wedge tool. This technique is known for producing shorter, more direct connections, which reduce parasitic capacitance and improve signal integrity.

Wedge bonding excels in applications that demand high-frequency performance and mechanical reliability [13]. It is often used in aerospace and military systems, RF and microwave devices, and power electronics, where it provides robust connections capable of withstanding vibration, temperature fluctuations, and shock. However, wedge bonding tends to be slower than ball bonding, which limits its use in high-volume production. It also requires precise alignment, especially in fine-pitch applications, making the process more challenging for intricate designs. Despite these limitations, the mechanical strength and reliability of wedge bonds make them a preferred choice for mission-critical systems. Differences of both the bonding's could be viewed in Fig. 2.2.

Aspect	Ball Bonding	Wedge Bonding
Bond Shape	Forms a **spherical bond** at the wire end.	Creates a **wedge-shaped bond** by flattening the wire.
Wire Material	Uses **gold (Au)** or **copper (Cu)** wires.	Typically employs **aluminum (Al)** or **gold (Au)** wires.
Bonding Speed	**Faster process**, suitable for high-volume production.	**Slower process** due to precision placement.
Applications	Common in **consumer electronics** like smartphones.	Preferred for **high-reliability applications** like aerospace and automotive systems.
Tool Movement	Requires **vertical movement** of the bonding tool.	Involves **horizontal tool motion**, increasing risk of misalignment.

Fig. 2.2 Comparison of ball bonding and wedge bonding techniques

2.2.3   Addressing Signal Integrity and High-Frequency Challenges

Both ball bonding and wedge bonding face challenges when it comes to signal interference and high-frequency applications. Longer wire paths, such as those formed during ball bonding, introduce inductance and capacitance, degrading signal quality and limiting the method's applicability in RF systems and high-speed data processing. Wedge bonding, with its shorter connections, mitigates some of these issues, making it ideal for microwave and high-frequency circuits. However, even with wedge bonding, achieving optimal signal integrity at higher frequencies requires advanced process optimization and material selection.

To overcome these challenges, manufacturers are employing several strategies. Shorter wire lengths are achieved through multi-bonding tools that minimize parasitics, while EMI shielding techniques reduce the impact of electromagnetic interference on sensitive circuits. Additionally, hybrid packaging approaches—combining wire bonding with flip-chip bonding or thermosonic bonding—enable devices to achieve both high-frequency performance and robust electrical connections [14].

2.3   Flip Chip Bonding: Evolution and Advantages Over Wire Bonding

Flip chip bonding is considered the most effective alternative to traditional wire bonding due to its ability to provide shorter interconnect lengths, higher I/O density, and improved electrical and thermal performance. The key distinguishing feature of flip chip packaging is the "flipped" IC, where the active side of the chip faces the substrate, allowing for direct connections through interconnects such as copper pillars, conductive adhesives, solder bumps, and stud bumps. This approach, known as Controlled Collapse Chip Connection (C4), was first introduced in the 1960s by IBM, which developed the method for packaging individual transistors and diodes used in mainframe systems [15]. C4 technology enabled the deposition of solder bumps on the active side of the semiconductor die, which were then aligned and bonded to corresponding pads on the package substrate. This arrangement significantly reduced the length of interconnects, thereby minimizing signal propagation delays and mitigating parasitic effects associated with longer wire bonds. The result was an improvement in both thermal and electrical performance, making flip chip bonding a compelling choice for high-performance applications.

Over the years, flip chip bonding has seen numerous advancements in materials, processes, and equipment, which have contributed to its growing popularity. The development of lead-free solders and improved solder pastes has enhanced the reliability and environmental sustainability of flip chip packages. Meanwhile, innovations such as copper pillar bumping have enabled finer pitch bonding and higher interconnect densities, which are essential for modern high-performance semiconductor packages. Copper pillars, in particular, offer better current-carrying capacity and thermal management compared to traditional solder bumps, making them suitable for devices with high power density.

Flip chip bonding is widely used in the packaging of CPUs [16], GPUs, and high-speed memory devices due to its productivity and capability to support high data processing rates. Unlike wire bonding, which connects only the periphery of the chip, flip chip bonding allows for bumps to be placed across the entire surface of the die. This capability increases the number of available inputs and outputs (I/O), enabling faster data transfer rates and more efficient power distribution. As a result, flip chip bonding is particularly well-suited for applications where high speed and high bandwidth are critical.

However, flip chip bonding is not without its limitations. One of the main challenges is the difficulty in achieving multi-chip stacking, which is often required for memory devices with high density requirements. Additionally, while flip chip bonding can accommodate more I/Os than wire bonding, the bump pitch and the pitch of the organic PCB can still limit the total number of I/Os that can be connected. To address these limitations, new packaging technologies such as TSV bonding have been developed. TSV bonding allows for vertical interconnections through the silicon wafer, enabling 3D stacking of chips, which increases I/O density and supports higher integration levels for memory and logic devices.

2.4 Micro Bump (C2) and Solder Bump (C4): Evolution and Applications (1980-Present)

The development of micro bump and solder bump technologies during the 1980s and 1990s marked a significant advancement in semiconductor packaging, driven by the increasing demand for higher interconnect densities and improved performance. These technologies, which include micro bumps (chip-to-chip or C2 connections) and solder bumps C4, revolutionized the ability to achieve fine-pitch bonding and high-density interconnects in advanced electronics [17].

Micro bump technology focuses on creating intricate, fine-pitch interconnects between semiconductor chips using metallized bumps or pillars. This approach allows for direct chip-to-chip integration, which has been instrumental in advancing multi-chip modules (MCMs) [18] and SoC solutions [19]. The smaller size and finer pitch of micro bumps facilitate a higher number of interconnections in a given area, making them particularly suitable for applications where miniaturization is critical. These include small form factor devices such as mobile phones, wearables, and IoT gadgets [20]. Micro bumps are essential for high-density interposers and 3D stacking applications, where they enable complex routing and integration of multiple layers in 3D ICs and 3D NAND memory [21]. The use of micro bumps also supports heterogeneous integration, allowing different types of components—such as analog, digital, and MEMS devices—to be combined within a single package.

Solder bumps, on the other hand, have a longer history in the industry and are widely used in high-performance computing, microprocessors, and other power-intensive applications.

C4 technology, which involves larger bumps compared to micro bumps, provides higher current-carrying capacity and superior thermal dissipation capabilities. This makes solder bumps particularly suitable for applications where power and heat management are critical, such as CPUs and GPUs. The larger size and volume of solder bumps enable them to handle higher current densities, making them advantageous for power-hungry devices. However, the larger pitch size associated with solder bumps limits the achievable interconnect density and scalability for miniaturization efforts, and they tend to incur higher manufacturing costs due to the amount of material used.

The micro bump bonding process involves several key steps, starting with the deposition of solder materials on fine-pitch pads using techniques such as electroplating. Precision alignment tools are then used to position the micro bumps accurately with the corresponding pads on the substrate or interposer. Finally, a controlled heating process, known as reflow, forms solid connections between the micro bumps and the receiving pads. While micro bumps provide advantages in terms of miniaturization, signal integrity, and compatibility with advanced packaging technologies, they also present challenges. Their smaller size and tighter tolerances increase the difficulty of achieving consistent alignment and bonding, potentially leading to issues with bump collapse during bonding and challenges in maintaining high manufacturing yields at finer pitches.

Despite these challenges, micro bumps remain a preferred choice for meeting the demands of compact, high-performance electronic devices. Their ability to provide high interconnect density while supporting emerging technologies like 2.5D interposers and 3D integrated circuits has solidified their role in modern semiconductor packaging. As both micro bump and solder bump technologies continue to evolve, ongoing advancements in materials, equipment, and processes are expected to further enhance their reliability, scalability, and overall performance in the ever-growing field of semiconductor packaging.

2.5   TSV Bonding: Revolutionizing 3D Integration (Early 2000-Present)

TSV bonding has transformed the semiconductor packaging industry by enabling 3D integration and Fig. 2.3. Illustrates the same, where multiple layers of circuits and components are vertically connected [22]. TSV technology eliminates the need for traditional horizontal wiring by drilling precise, vertical pathways through the silicon wafer, allowing for direct, low-latency communication between stacked chips, Fig. 2.4 depicts the clear process flow of TSV fabrication flow for 3D Integration. Each via is filled with a conductive material, such as copper, to form electrical connections between layers. TSV bonding has become a cornerstone of advanced packaging, addressing key challenges in miniaturization, performance optimization, and power efficiency, and is now used in high-performance applications ranging from memory architectures to logic processors.

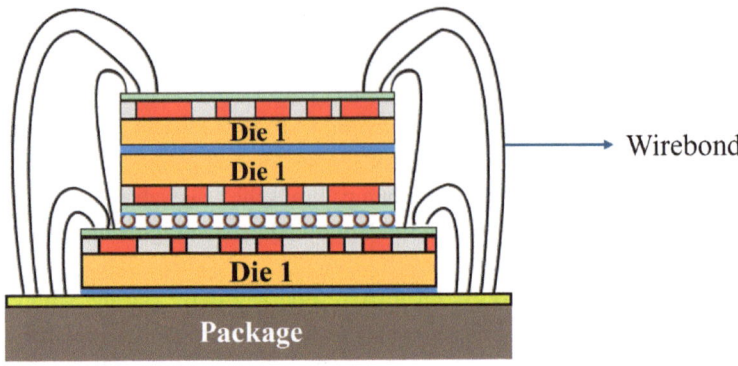

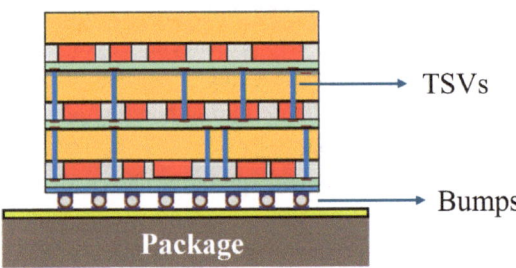

Fig. 2.3 TSV application in 3D ICs showcasing traditional wire-bonded 3D packages and TSV interconnected dies

Introduced in the early 2000s, TSV technology initially found applications in memory devices and CMOS image sensors, where it significantly increased device density and improved performance. Over time, advancements in fabrication processes such as deep reactive ion etching (DRIE) and TSV-last techniques [23] expanded its applicability to a wider range of products, including microprocessors, SoC solutions, and graphics processors. These innovations enabled the creation of smaller, more powerful devices, with TSV bonding facilitating the integration of multiple dies in a single package, reducing form factors, and enhancing system-level performance.

Application of TSV in Advanced Packaging TSV technology plays a vital role in several high-performance and space-constrained applications across the semiconductor industry. Its ability to stack and integrate multiple dies vertically has opened new possibilities for advanced packaging architectures.

High Bandwidth Memory (HBM) and 3D DRAM: TSVs are extensively used in HBM and 3D DRAM stacks, where they connect multiple layers of memory to achieve higher

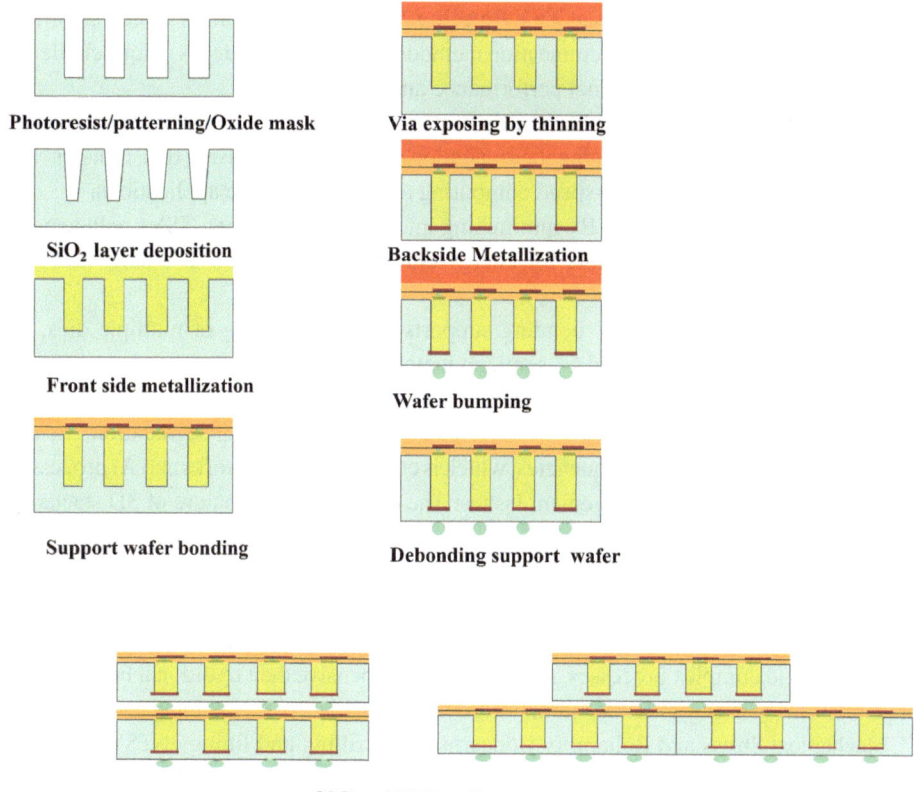

Photoresist/patterning/Oxide mask

SiO₂ layer deposition

Front side metallization

Support wafer bonding

Via exposing by thinning

Backside Metallization

Wafer bumping

Debonding support wafer

C2C or C2W bonding

Fig. 2.4 TSV fabrication process flow for 3D integration

data transfer rates. These architectures support high-performance workloads in AI, cloud computing, and gaming, where speed and efficiency are paramount.

Logic-Memory Integration in Processors: TSVs enable tight integration of logic and memory layers, reducing communication latency between them. This configuration is essential in CPUs, GPUs, and SoCs, particularly for data center applications and AI accelerators, where rapid data access is crucial [24]

CMOS Image Sensors and Advanced Camera Modules: TSVs are employed in CMOS image sensors to connect pixel arrays with readout circuits, enabling higher resolutions and faster frame rates. These sensors are critical in smartphones, medical imaging, and autonomous vehicles, where precision and speed are essential [25].

3D NAND Memory and Edge Devices: TSVs also support 3D NAND memory architectures, providing the density and bandwidth needed for modern storage solutions. Additionally, they are used in edge devices that require compact, energy-efficient designs with high processing power.

Advantages and Challenges of TSV Bonding TSV technology offers several significant advantages over traditional interconnection methods, but it also introduces unique challenges that must be addressed for optimal performance and reliability.

Advantages: High Bandwidth and Low Latency: TSVs enable shorter electrical paths between components, reducing signal propagation delays and increasing data transfer rates. This makes them ideal for high-speed computing and data-intensive applications.

Lower Power Consumption: By minimizing interconnect distances, TSVs reduce power dissipation, making them suitable for energy-efficient devices such as smartphones and wearable electronics.

Compact Form Factor: TSV bonding supports vertical stacking of multiple dies, significantly reducing the footprint of semiconductor packages. This is essential for space-constrained devices, where miniaturization is a priority.

Improved Signal Integrity: Shorter interconnect paths reduce signal degradation, ensuring better performance at higher frequencies, which is critical for 5G networks and AI processors.

Challenges: Thermal Management Issues: The densely packed nature of 3D stacks creates challenges in dissipating heat. Hotspots can form within the stacked layers, potentially affecting performance and reliability. Advanced cooling techniques, such as thermal interface materials and microfluidic cooling, are necessary to mitigate these issues.

Manufacturing Complexity and Cost: TSV bonding requires precise alignment, advanced equipment, and complex processes, making it more expensive than traditional bonding techniques. Ensuring high yields and minimizing defects is critical to maintaining cost-efficiency.

Mechanical Stress and Reliability Concerns: The drilling and filling of TSVs introduce mechanical stress into the silicon, which can impact long-term reliability. Stress-relief techniques, such as redundant vias and stress buffers, are used to mitigate these effects.

Defect Detection and Quality Control: Ensuring the integrity of TSVs throughout the manufacturing process is challenging. Non-destructive inspection methods, such as X-ray imaging and acoustic microscopy, are employed to identify defects and ensure bond reliability [26].

2.6   Thermocompression Bonding: Precision Through Heat and Pressure

Thermocompression bonding combines heat, pressure, and time to create durable interconnects by softening the bonding material—often gold or copper—to adhere it to a contact surface [27]. In this process, the materials to be bonded are carefully aligned, and controlled pressure and heat are applied over a set period to achieve a strong, reliable bond. This technique is particularly suited for fine-pitch applications and high-density interconnects, where precise alignment is essential.

Thermocompression bonding is used in microprocessors, logic devices, and 3D packaging due to its ability to handle narrow pitches and high aspect ratios. However, achieving successful bonds requires precise surface preparation and alignment, as misalignment or contamination can cause defects. The application of sufficient pressure and temperature ensures that the materials adhere without requiring additional adhesives, which reduces resistance and improves electrical performance.

While thermocompression bonding is ideal for high-performance devices, it presents challenges related to thermal management and potential warping [28] of substrates due to the heat applied during bonding. Despite these hurdles, its ability to enable high-density integration makes it a preferred choice in advanced semiconductor packaging.

2.7   Thermosonic Bonding: Ultrasonic Energy for Delicate Components

Thermosonic bonding combines ultrasonic energy and heat to bond components, typically gold wires to delicate surfaces such as MEMS devices, sensors, and RF components. This technique offers low bonding force, making it ideal for fragile components that cannot withstand high pressure or temperatures. During the bonding process, ultrasonic vibrations are applied through a bonding tool to generate localized heat and energy, softening the gold wire and forming a secure connection with the bond pad.

Thermosonic bonding is especially valuable in applications where high mechanical strength is not the primary requirement but gentle bonding is essential to avoid damaging the device. It is commonly used in medical sensors, RF modules, and aerospace applications, where reliability and precision are paramount. Although slower than other bonding methods like ball bonding, thermosonic bonding ensures strong connections with minimal risk of damage, extending the lifespan of delicate devices.

2.8   Anisotropic Conductive Film (ACF) Bonding

A Solution for Flexible Electronics Anisotropic Conductive Film (ACF) bonding is a specialized technique that uses a polymer film embedded with conductive particles to connect two components [29]. The ACF is placed between the components, and pressure and heat are applied to activate the adhesive and align the conductive particles, creating electrical pathways only in specific directions (anisotropically). This selective conductivity makes ACF bonding ideal for fine-pitch applications that require both mechanical flexibility and electrical performance.

ACF bonding plays a crucial role in the manufacturing of flexible displays, wearable electronics, and biosensors, where circuits need to endure bending, stretching, and repeated mechanical deformation. The polymer matrix of the ACF provides mechanical stability, while the embedded particles ensure reliable electrical connections [30]. However, achieving uniform alignment of the conductive particles is critical, as improper bonding can lead to resistance variations or open circuits. Advances in material science are driving improvements in ACF performance, making it a preferred technology for next-generation flexible and wearable devices.

2.9   Hybrid Bonding: The Backbone of 3D ICs and Heterogeneous Integration (2020-Present)

Hybrid bonding, also referred to as direct bonding, represents a breakthrough technology in 3D ICs [31]. It involves the direct contact between metal pads and dielectric surfaces without the need for adhesives or solder, enabling a more seamless and efficient bonding process. Hybrid bonding allows for wafer-to-wafer or die-to-wafer bonding, making it crucial for high-density vertical integration and heterogeneous packaging.

In hybrid bonding, surfaces are prepared to be atomically smooth and clean to ensure perfect adhesion. Once aligned, the metal pads and dielectric materials bond through inter-molecular forces, resulting in an interconnect with low resistance and high electrical performance. The absence of solder or adhesive improves the thermal and mechanical properties of the interconnect, making hybrid bonding ideal for high-performance devices such as processors, AI accelerators, and memory stacks.

Oxide-Metal Hybrid Bonding involves bonding a metal surface to an oxide surface, leveraging direct metal-to-dielectric contact. This technique enhances the bond's electrical and thermal conductivity, as well as its mechanical stability. Oxide-metal bonding is particularly advantageous in high-performance applications, as it provides low-resistance interconnections and improved heat dissipation, making it ideal for 3D IC stacks where thermal management is critical.

Polymer-Metal Hybrid Bonding combines metal pads with polymer-based dielectrics, offering flexibility and lower processing temperatures compared to oxide bonding. This technique is useful in applications requiring some degree of mechanical compliance, such as flexible electronics or chip packaging that may undergo bending or deformation. Polymer-metal bonding also helps reduce stress between bonded layers, improving the overall reliability of devices in applications like flexible displays or sensor arrays. Both of these Hybrid bonding methods are illustrated in Fig. 2.5.

Hybrid bonding as depicted in Fig. 2.6. is increasingly used in chiplet-based architectures [32], where different types of chips—such as logic, memory, and sensors—are integrated into a single package. It enables faster data transmission, better power management, and smaller form factors, all of which are essential for advanced packaging technologies. However, the process requires high precision and stringent surface preparation, making it more challenging to implement compared to traditional bonding methods.

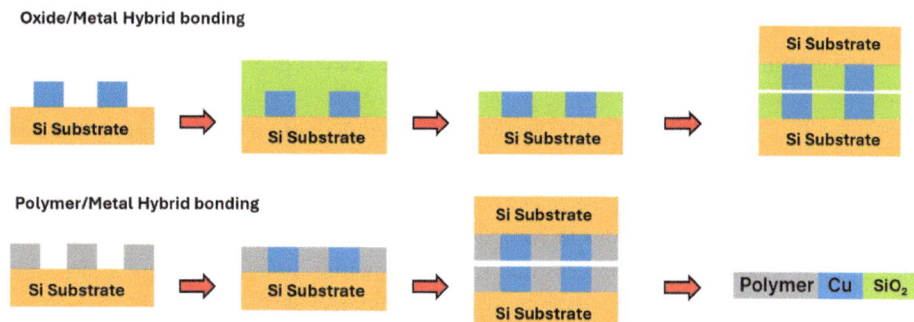

Fig. 2.5 Process flow of various hybrid bonding techniques, including oxide-metal hybrid bonding and polymer-metal hybrid bonding

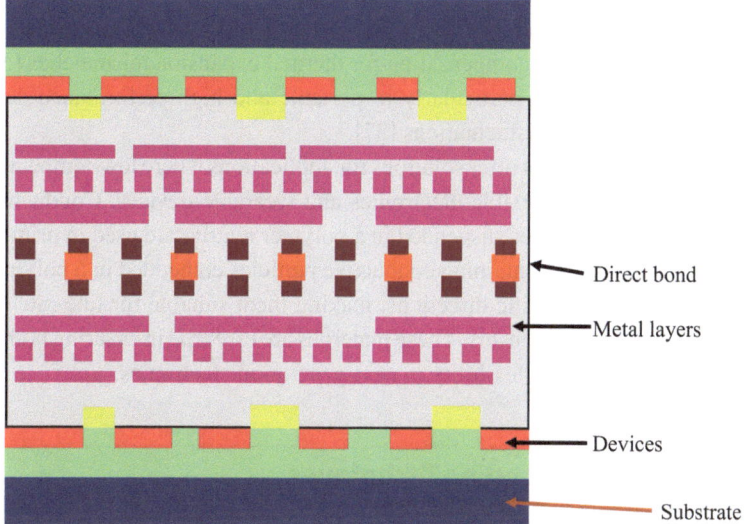

Fig. 2.6 Illustration of hybrid bonding (Direct Bonding) mechanism

2.10   Materials for Bonding Interconnects

Bonding techniques in semiconductor packaging rely on a wide range of materials to ensure robust, reliable, and efficient interconnects [33]. Metals such as gold, copper, and aluminum are commonly used due to their excellent electrical conductivity and mechanical strength. Gold, known for its superior corrosion resistance, is frequently employed in high-reliability applications, including aerospace and medical devices. Copper, on the other hand, offers better conductivity and is preferred in flip-chip bonding and thermocompression bonding for high-performance computing, although it requires careful handling to prevent oxidation.

Aluminum, often used in wedge bonding, is valued for its low cost and compatibility with low-temperature processes, making it ideal for power electronics and automotive applications.

Solder alloys are another essential category of materials, particularly in flip-chip bonding, where small solder bumps are reflowed to create electrical connections. Traditional tin-lead (Sn-Pb) alloys, once the industry standard, are being replaced by lead-free alternatives such as tin-silver-copper (SAC) alloys to comply with environmental regulations. These alloys, though more challenging to process due to higher melting points, provide the necessary mechanical strength and thermal stability for advanced packaging solutions [34].

In addition to metals and solder, adhesives and polymers play a critical role in maintaining the mechanical integrity of bonded assemblies. Epoxy adhesives are widely used to bond non-metallic components and fill gaps between materials, offering resistance to thermal cycling and environmental stresses [35]. Underfill materials, applied between the chip and substrate in flip-chip packages, provide additional support by absorbing mechanical stress, preventing delamination, and compensating for thermal expansion mismatches [36]. These polymers enhance reliability, particularly in portable and high-performance devices that undergo frequent temperature fluctuations [37].

Conductive pastes and films offer solutions for applications where low-temperature bonding is required, such as in flexible electronics and wearable devices. Conductive pastes, which consist of metal particles suspended in a polymer matrix, are used in printed circuits and biosensors [38]. ACFs, containing conductive particles embedded in a polymer, enable selective conductivity in specific directions, making them suitable for fine-pitch interconnects in flexible displays and sensors. These materials are essential in ensuring both electrical performance and mechanical flexibility in next-generation electronics.

2.11   Challenges in Bonding Techniques

While bonding techniques are crucial for creating reliable interconnects, they come with several challenges, particularly as devices become smaller and more complex. One of the primary challenges is achieving precise alignment. As interconnect density increases and pitch scales down to a few microns, even minor misalignments can lead to short circuits, open connections, or signal interference, reducing yield and performance. This is especially critical in thermocompression and hybrid bonding, where high precision is required to align components accurately.

Reliability is another key concern. Bonded interconnects are subject to fatigue and cracking over time due to mechanical stress and thermal cycling. These stresses can cause failures, particularly in applications like automotive electronics and aerospace systems, where devices experience constant vibrations and temperature fluctuations. Electromigration, the gradual movement of metal atoms under high current density, presents another risk. This phenomenon can result in void formation within the interconnects, compromising electrical continuity and leading to device failure, particularly in copper-based interconnects [39].

Thermal stress poses additional challenges, especially in flip-chip packaging, where the silicon die and organic substrate have different thermal expansion coefficients. Repeated heating and cooling cycles can cause delamination or cracks in the bonded layers. Underfill materials are often used to mitigate this stress, but precise control of bonding parameters—such as temperature, pressure, and bonding time—is essential to avoid thermal damage during the bonding process.

Material compatibility is another significant issue, particularly in heterogeneous integration, where different types of chips—such as logic, memory, and sensors—are combined within a single package. Differences in surface properties, thermal expansion, and chemical composition can impact bonding quality. Advanced bonding techniques, such as hybrid bonding, address some of these compatibility challenges by requiring atomically smooth surfaces to ensure strong, reliable connections. However, maintaining such stringent surface conditions adds complexity to the bonding process.

In summary, while bonding techniques provide the foundation for reliable interconnects, they must overcome challenges related to alignment accuracy, thermal management, electromigration, and material compatibility. As semiconductor packaging continues to evolve toward smaller, more integrated systems, innovations in bonding technologies will be essential to ensure long-term performance, reliability, and scalability in advanced microelectronics.

2.12 Industry Leaders and the Bonding Ecosystem

The bonding ecosystem in semiconductor manufacturing is a complex, collaborative network comprising chip manufacturers, packaging service providers, equipment vendors, and component suppliers. As advanced bonding technologies continue to evolve, leading companies in the industry drive innovation, ensuring that modern devices meet the demands of performance, reliability, and miniaturization. This section explores the roles of key industry leaders and equipment vendors, highlighting their contributions to bonding processes in high-volume production and R&D applications.

Major Industry Leaders: Chipmakers and Packaging Providers Several companies dominate the field of semiconductor packaging and bonding technologies, providing cutting-edge solutions for both 3D integration and heterogeneous packaging. These companies invest heavily in R&D to push the boundaries of interconnect density, power efficiency, and reliability.

Amkor Technology: Amkor is one of the world's largest OSAT providers. The company offers a wide range of advanced packaging solutions, including flip-chip technology, WLP, SiP, and 3D IC integration. Amkor's expertise in thermocompression bonding and hybrid bonding makes it a key player in delivering high-performance interconnect solutions for automotive, consumer electronics, and AI markets [40].

ASE Group: ASE Group is a leader in chiplet-based packaging and heterogeneous integration. Known for its focus on 3D packaging, FOWLP, and flip-chip bonding, ASE plays a crucial role in enabling advanced interconnects for AI accelerators, 5G, and IoT devices. The company's strong partnerships with chipmakers and equipment providers allow it to develop scalable and innovative bonding solutions.

Intel Corporation: Intel is at the forefront of chiplet-based architectures and heterogeneous integration. Through its advanced EMIB and Foveros 3D stacking technologies, Intel leverages thermocompression and hybrid bonding to create efficient, high-density interconnects. These technologies support Intel's goal of scaling performance in data centers, PCs, and edge computing. Intel's emphasis on AI-powered automation in bonding processes enhances both precision and yield.

Taiwan Semiconductor Manufacturing Company (TSMC): TSMC is a pioneer in 3D ICs and advanced packaging, with significant investments in WLP and TSVs. TSMC's CoWoS (Chip on Wafer on Substrate) and InFO (Integrated Fan-Out) technologies rely on sophisticated bonding techniques to achieve low-latency interconnects. These packaging solutions are widely adopted in mobile devices, high-performance computing (HPC), and AI accelerators.

Samsung Electronics: Samsung is known for its leadership in memory technologies and logic ICs. The company uses advanced flip-chip bonding and hybrid bonding to create efficient interconnects for HBM and processors in smartphones and servers. Samsung's innovations in 3D IC integration and wafer-level packaging position it as a key player in next-generation semiconductor technologies.

2.13 Equipment Vendors: Supporting Bonding Processes in High-Volume Production and R&D

To meet the growing demands of advanced bonding techniques, specialized equipment manufacturers provide the tools necessary for high-precision bonding, alignment, and automation. These equipment vendors support both large-scale production and research environments, ensuring that new technologies are scalable and reliable.

Kulicke and Soffa: Kulicke and Soffa is a global leader in wire bonding equipment and flip-chip solutions. The company offers a range of tools for ball bonding, wedge bonding, and thermosonic bonding, widely used in the automotive, industrial, and consumer electronics sectors. Its bonding machines feature AI-powered alignment systems that improve precision and reduce process variability in fine-pitch applications.

Besi (BE Semiconductor Industries): Besi specializes in die attach and flip-chip bonding equipment for high-end semiconductor packaging. The company's tools are designed for thermocompression bonding, hybrid bonding, and wafer-level packaging. Besi's systems support 3D IC stacking and chiplet-based architectures, enabling the production of compact, high-performance devices used in data centers, smartphones, and wearable electronics.

ASM Pacific Technology: ASM Pacific provides a wide range of bonding and assembly equipment, including tools for wire bonding, flip-chip bonding, and thermocompression bonding. ASM's solutions are designed to meet the requirements of high-volume manufacturing while maintaining the flexibility needed for R&D environments. Its machines are widely used in MEMS packaging, RF devices, and power electronics.

Shibaura Machine: Shibaura offers advanced die bonding and flip-chip assembly equipment. The company's tools are known for their high precision and reliability, making them ideal for wafer-level packaging and 3D IC integration. Shibaura's systems are widely adopted in semiconductor fabs across Asia and play a critical role in high-performance packaging solutions.

The Role of R and D Institutions and Component Suppliers In addition to chipmakers and equipment vendors, research institutions and component suppliers contribute to the development of advanced bonding technologies. Universities and research centers collaborate with the industry to explore new bonding materials, AI-based alignment tools, and sustainable processes. Innovations from these collaborations help the industry address challenges such as thermal management, material compatibility, and process scalability.

Component suppliers also play a vital role by providing critical parts for bonding equipment, such as vacuum chambers, power supplies, alignment tools, and conductive pastes. Companies like Advanced Energy, ANCORP, and VacTechniche supply essential components that ensure the smooth operation of bonding systems in both production facilities and R&D labs.

Conclusion:
This chapter provided a comprehensive exploration of bonding techniques and their significance in semiconductor manufacturing and advanced packaging. We examined various bonding methods, including wire bonding, thermocompression, thermosonic, ACF, and hybrid bonding, each suited for specific applications and performance needs. The discussion highlighted the role of bonding in enabling high-density interconnects, chiplet architectures, and 3D integration. Critical materials like gold, copper, solder alloys, and polymers were analyzed, along with the challenges posed by alignment, fatigue, thermal stress, and material compatibility. Industry leaders such as Amkor, ASE Group, Intel, and TSMC, supported by equipment vendors like Kulicke and Soffa and Besi, continue to drive innovation, ensuring bonding technologies evolve to meet the demands of next-generation microelectronics.

References

1. Charles, Harry K. "Advanced wire bonding technology: materials, methods, and testing." *Materials for Advanced Packaging* (2009): 113–79.
2. Zulkifli, Muhammad Nubli, et al. "Some thoughts on bondability and strength of gold wire bonding." *Gold Bulletin* 45 (2012): 115–125.
3. Schneider-Ramelow, Martin, and Christian Ehrhardt. "The reliability of wire bonding using Ag and Al." *Microelectronics Reliability* 63 (2016): 336–341.
4. Tok, C. W., et al. "Wire bonding improvement through optimal bonding tools and materials selection." In *2007 9th Electronics Packaging Technology Conference*. IEEE, 2007.
5. Charles, Harry K. "Advanced wire bonding technology: Materials, methods, and testing." *Materials for Advanced Packaging* (2017): 131–198.
6. Kim, Jongbaeg, et al. "Ultrasonic bonding for MEMS sealing and packaging." *IEEE Transactions on Advanced Packaging* 32.2 (2009): 461–467.
7. Lim, Michael Joo Zhong, et al. "Copper wire degradation under the effect of different humidity levels." In *2023 IEEE 73rd Electronic Components and Technology Conference (ECTC)*. IEEE, 2023.
8. Hung, Tzu-Heng, Yu-Ming Pan, and Kuan-Neng Chen. "Stress issue of vertical connections in 3D integration for high-bandwidth memory applications." *Memories-Materials, Devices, Circuits and Systems* 4 (2023): 100024.
9. Gan, Chong Leong, et al. "Extended reliability of gold and copper ball bonds in microelectronic packaging." *Gold Bulletin* 46 (2013): 103–115.
10. Hsueh, Hao-Wen, et al. "Microstructure, electric flame-off characteristics and tensile properties of silver bonding wires." *Microelectronics Reliability* 51.12 (2011): 2243–2249.
11. Jacob, Peter, and Michael Rütsch. "Stitch-Bond-Shearing in Optoelectronic Devices, Caused by Lead-Free-Wave-Soldering–Do We Need Improved Wire-Bonding Methods?" In *International Symposium for Testing and Failure Analysis*. Vol. 30910. ASM International, 2008.
12. Supramaniam, Saraswathy, et al. "Moisture Content and Early Corrosion Detection of Cu Wire Bonding in a Semiconductor Package." *Journal of Failure Analysis and Prevention* 23.6 (2023): 2362–2369.
13. Wei, Xing, et al. "Reliability evaluation of thick Ag wire bonding on Ni pad for power devices." *Microelectronics Reliability* 152 (2024): 115304.
14. Wu, Yongshuan, et al. "Process Parameters and Modeling Study of Thermosonic Flip Chip Bonding." *IEEE Transactions on Components, Packaging and Manufacturing Technology* 13.4 (2023): 545–552.
15. Wei, C. C., et al. "Comparison of the electromigration behaviors between micro-bumps and C4 solder bumps." In *2011 IEEE 61st Electronic Components and Technology Conference (ECTC)*. IEEE, 2011.
16. Underwood, Keith. "FPGAs vs. CPUs: trends in peak floating-point performance." In *Proceedings of the 2004 ACM/SIGDA 12th International Symposium on Field Programmable Gate Arrays*. 2004.
17. Lee, Chang-Chun, Tzai-Liang Tzeng, and Pei-Chen Huang. "Development of simulation-approach for 3D chip stacking with fine-pitch array-type microbumps." In *2015 International Conference on Electronics Packaging and iMAPS All Asia Conference (ICEP-IAAC)*. IEEE, 2015.
18. Ho, David, et al. "Chiplet Solution with FO-MCM Package in Edge and Cloud Computing (IMPACT 2023)." In *2023 18th International Microsystems, Packaging, Assembly and Circuits Technology Conference (IMPACT)*. IEEE, 2023.
19. Chang, Henry, et al. *Surviving the SoC Revolution*. Dordrecht: Kluwer Academic Publishers, 1999.

20. Allioui, Hanane, and Youssef Mourdi. "Exploring the full potentials of IoT for better financial growth and stability: A comprehensive survey." *Sensors* 23.19 (2023): 8015.
21. Shim, Sun Il, Jaehoon Jang, and Jaihyuk Song. "Trends and future challenges of 3D NAND flash memory." In *2023 IEEE International Memory Workshop (IMW)*. IEEE, 2023.
22. Wang, Jintao, et al. "A short review of through-silicon via (TSV) interconnects: metrology and analysis." *Applied Sciences* 13.14 (2023): 8301.
23. Shen, Jiayi, et al. "Impact of Super-long-throw PVD on TSV Metallization and Die-to-Wafer 3D Integration Based on Via-last." In *2023 IEEE International 3D Systems Integration Conference (3DIC)*. IEEE, 2023.
24. Chatterjee, Swetaki, et al. "Ferroelectric FDSOI FET modeling for memory and logic applications." *Solid-State Electronics* 200 (2023): 108554.
25. Wang, Bozhi, et al. "All-dielectric metasurface-based color filter in CMOS image sensor." *Optics Communications* 540 (2023): 129485.
26. Kumari, Vandana, et al. "Reliability Concerns of TSV based 3D Integration: Impact of Interfacial Crack." *IEEE Transactions on Components, Packaging and Manufacturing Technology* (2023).
27. Huang, Yuan-Chiu, et al. "Cu-Based Thermocompression Bonding and Cu/Dielectric Hybrid Bonding for Three-Dimensional Integrated Circuits (3D ICs) Application." *Nanomaterials* 13.17 (2023): 2490.
28. Alzyod, Hussein, and Peter Ficzere. "Thermal Evaluation of Material Extrusion Process Parameters and Their Impact on Warping Deformation." *Jordan Journal of Mechanical & Industrial Engineering* 17.4 (2023).
29. Qiao, Pengyi, et al. "Research on the related properties of ACF anisotropic conductive adhesive." *Highlights in Science, Engineering and Technology* 56 (2023): 308–314.
30. Xu, Yadong, et al. "Structural Analysis of Anisotropic Conductive Film for Liquid Crystal Displays and Semiconductor Packaging Applications." In *2023 24th International Conference on Electronic Packaging Technology (ICEPT)*. IEEE, 2023.
31. Cheemalamarri, Hemanth Kumar, et al. "Cu/Dielectric Hybrid Bonding Among Glass and Si." In *2024 IEEE 74th Electronic Components and Technology Conference (ECTC)*. IEEE, 2024.
32. Sudarshan, Chetan Choppali, et al. "ECO-CHIP: Estimation of Carbon Footprint of Chiplet-based Architectures for Sustainable VLSI." In *2024 IEEE International Symposium on High-Performance Computer Architecture (HPCA)*. IEEE, 2024.
33. Moon, Jun Hwan, et al. "Materials quest for advanced interconnect metallization in integrated circuits." *Advanced Science* 10.23 (2023): 2207321.
34. Dele-Afolabi, T. T., et al. "Recent advances in Sn-based lead free solder interconnects for microelectronics packaging: materials and technologies." *Journal of Materials Research and Technology* (2023).
35. Wang, Ruikun, et al. "A study of the residual stress behavior of rigid and flexible epoxy adhesives during thermal cycle aging for electronics packaging." *Journal of Adhesion Science and Technology* 38.4 (2024): 517–532.
36. He, Tao, et al. "A modified Qian-Liu constitutive model based on advanced packaging underfill materials." *Materials Science in Semiconductor Processing* 176 (2024): 108340.
37. Huang, Hua-Dong, et al. "Promising strategies and new opportunities for high barrier polymer packaging films." *Progress in Polymer Science* (2023): 101722.
38. Zhang, Bowen, et al. "Development of Silver Paste With High Sintering Driving Force for Reliable Packaging of Power Electronics." *IEEE Transactions on Components, Packaging and Manufacturing Technology* (2024).
39. Shen, Zesheng, et al. "Electromigration in three-dimensional integrated circuits." *Applied Physics Reviews* 10.2 (2023).
40. Choi, Byung Hun. "A Study on Competitiveness of Taiwan's Semiconductor Packaging & Testing Business." <processinggraph/> 37 (2023): 105–133.

CVD in Semiconductor Packaging

3

3.1 An Introduction to the World of Chemical Vapor Deposition

CVD stands as a cornerstone technology in the realm of semiconductor manufacturing, offering unparalleled capabilities in the deposition of materials crucial for the development of advanced semiconductor devices [1]. As an integral part of the fabrication process, CVD enables the precise application of thin films that are fundamental to device functionality and overall integrity. This chapter aims to explore the intricate principles of CVD, highlight its diverse applications within the field of semiconductor packaging, and shed light on the ongoing technological innovations that continue to shape its evolution.

CVD processes are essential for creating layers of materials atomically tailored to meet specific electronic functions, ranging from electrical insulation and thermal management to the formation of conductive pathways and optical interfaces. The versatility of CVD lies in its ability to deposit a wide range of materials with high precision over complex substrate topographies, making it indispensable for modern microelectronics. The process involves introducing one or more volatile precursors [2] into a reaction chamber [3], where they decompose or react at elevated temperatures on a substrate, forming a solid material as a thin film [4].

This deposition technique is celebrated for its scalability and ability to conformally coat [5] substrates with complex shapes, which is critical in applications such as TSVs for 3D integrated circuits, protective coatings for wearables, and photonic devices. Each application leverages the unique ability of CVD to control film properties at a microscopic level, thus ensuring the performance requirements of highly integrated circuits are met. For instance, in advanced logic and memory devices, CVD is used to deposit dielectric materials that provide electrical isolation between metal interconnects, thereby enhancing device speed and reducing power consumption.

The evolution of CVD technology has been closely tied to advancements in semiconductor materials science. Innovations in CVD equipment and techniques have enabled the

© The Author(s), under exclusive license to Springer Nature Switzerland AG 2025   41
N. Asadizanjani et al., *Introduction to Microelectronics Advanced Packaging Assurance*,
Synthesis Lectures on Engineering, Science, and Technology,
https://doi.org/10.1007/978-3-031-86102-4_3

deposition of films at lower temperatures, reduced chemical waste, and increased the types of materials that can be deposited. This adaptability makes CVD a key player in addressing the industry's push towards smaller, faster, and more energy-efficient devices. For example, the development of low-k dielectrics via CVD has been pivotal in reducing capacitance and power dissipation in high-performance ICs [6].

Moreover, the drive for greater functionality packed into smaller dimensions has spurred the use of ALD [7], a derivative of CVD that offers atomic-level thickness control. ALD's ability to coat high-aspect-ratio structures uniformly has made it critical for emerging applications, such as the fabrication of TSVs and advanced memory architectures. This nuanced control is crucial for maintaining the structural integrity and connectivity of microscale components.

As the semiconductor industry continues to evolve towards complex device architectures, including heterogeneous integration and flexible electronics, the role of CVD is expected to expand even further. Ongoing research and development are directed at enhancing the deposition rates, material quality, and environmental sustainability of CVD processes. The future of CVD is seen not only as a facilitator of current manufacturing needs but also as a driver of new material discoveries and applications that could redefine what is possible in semiconductor technology.

3.2   Principles of CVD

3.2.1   Basic Mechanism

At its core, CVD involves the deposition of a solid material from a gaseous phase onto a substrate through chemical reactions at elevated temperatures. This process is highly favored in semiconductor manufacturing for its ability to produce uniform, high-quality films that are essential for the intricate architectures of modern microelectronics.

Chemical Reaction Dynamics: In a typical CVD process, precursor gases containing the desired film constituents are introduced into a reaction chamber where they are thermally activated. Upon reaching the heated surface of the substrate, these gases undergo a series of complex chemical reactions. These reactions can result in the decomposition or reduction of the precursor gases, leading to the deposition of a solid film on the substrate while by-products are expelled as gases. The dynamics of these reactions are influenced by a variety of factors including the chemical properties of the precursors, the surface condition of the substrate, and the ambient conditions within the reactor. For instance, the deposition of silicon dioxide might involve the reaction of silane (SiH4) with oxygen:

$$SiH_4(g) + 2O_2(g) \rightarrow SiO_2(s) + 2H_2O(g)$$

This reaction is exothermic and needs careful thermal management to ensure film quality and prevent unwanted secondary reactions.

Deposition Techniques: CVD can be performed using a variety of techniques such as [8] atmospheric pressure CVD (APCVD), [9] low-pressure CVD (LPCVD), and [10] plasma-enhanced CVD (PECVD), each suited to specific applications based on the required material properties and device specifications. LPCVD, for example, is preferred for its high uniformity and material quality, crucial for manufacturing layers with minimal defects.

3.2.2 Key Parameters

The quality, uniformity, and electrical properties of CVD films are significantly influenced by process parameters such as temperature, pressure, and gas flow. Understanding and controlling these parameters is crucial for achieving desired film characteristics.

Temperature is a critical factor in CVD processes, as it determines the rate of chemical reactions and the quality of the film growth. High temperatures typically increase the reaction rates and improve the crystallinity of the deposited films. However, excessively high temperatures may also degrade sensitive substrates or result in unwanted diffusion of materials.For example, the deposition of epitaxial silicon layers [11] is often performed at temperatures around $1000\,°C$ to ensure the film grows with the correct crystal orientation and minimal defects. The temperature needs to be precisely controlled to align the deposition rate with the desired film thickness and purity.

Pressure within the CVD reactor influences the mean free path of the gas molecules, affecting both the reaction rate and the uniformity of the film deposition. Lower pressures are generally used in LPCVD systems to reduce gas-phase reactions and improve film uniformity across large substrates. Lower pressures help in reducing collisions among gas molecules, which in turn minimizes the likelihood of unwanted gas-phase nucleation—a common issue that can lead to particle defects in films [12].

Gas Flow The flow rate of the gases not only impacts the thickness and uniformity of the deposited films but also determines the deposition rate [13]. Optimal gas flow rates are essential to ensure a steady supply of reactants and the efficient removal of by-products. Controlled gas flow helps maintain a stable reaction zone and consistent material properties across the substrate. For instance, in the deposition of tungsten from WF6, adjusting the flow of hydrogen gas can selectively reduce WF6 to metallic tungsten, tightly controlling the deposition rate and film characteristics [14].

3.3   Types of CVD Processes

Illustrated in Fig. 3.1.

3.3.1   Low-Pressure CVD (LPCVD)

Mechanism and Benefits:
LPCVD is characterized by its operation under significantly reduced atmospheric pressure, which reduces the density of the gas molecules and minimizes collisions that could lead to unwanted secondary reactions. This setup extends the mean free path of the molecules, promoting more uniform surface reactions and higher quality film deposition.

The key benefit of LPCVD is the excellent uniformity and purity of the films it produces. By operating at lower pressures, LPCVD avoids the particle contamination often associated with higher pressure systems, leading to smoother and more defect-free surfaces. This process is particularly adept at creating stable, uniform layers of materials like silicon nitride and polysilicon across large wafers [15], which are crucial for consistent semiconductor device performance.

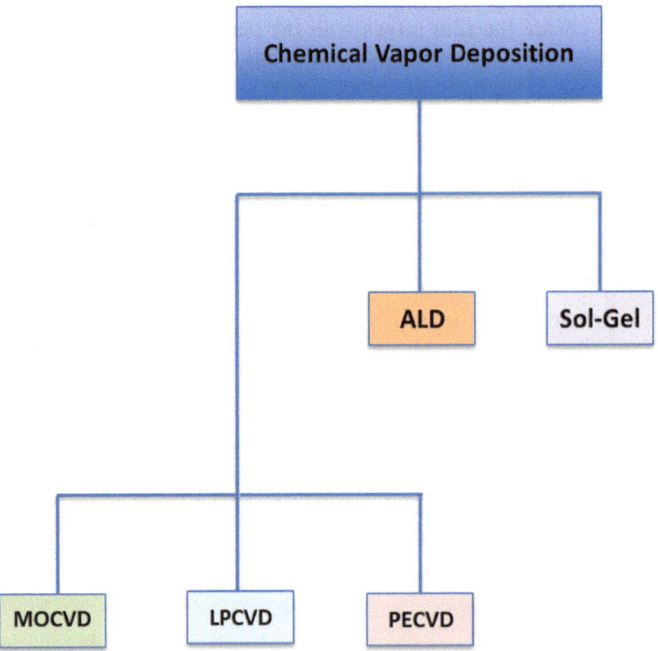

Fig. 3.1 Various types of chemical vapor deposition (CVD) techniques

3.3.2 Plasma-Enhanced CVD (PECVD)

Mechanism and Benefits:
PECVD incorporates plasma to reduce the thermal budget required for chemical reactions to occur [16]. By applying an electric field to the gas mixture, PECVD generates plasma, which consists of highly reactive species that facilitate film growth at temperatures lower than those typically required in conventional CVD processes.

PECVD's major advantage is its ability to deposit quality films on substrates that are thermally sensitive, making it suitable for a wide range of materials and applications [17]. The use of plasma enables enhanced control over film properties like density and stress, which are crucial for the reliability and performance of electronic devices. Moreover, PECVD can achieve excellent step coverage and uniformity on complex structures, supporting advanced microfabrication techniques.

3.3.3 Metal-Organic CVD (MOCVD)

Mechanism and Benefits:
MOCVD utilizes volatile metal-organic precursors to deposit thin films and its deposition process can be visualized through Fig. 3.2, primarily of compound semiconductors [18]. The process involves heating these precursors to decompose them into the desired materials, which then settle on the heated substrate.

The primary advantage of MOCVD lies in its ability to produce high-quality epitaxial films with precise control over composition and thickness [19]. It is particularly effective for creating complex multilayer structures with excellent crystalline quality, essential for optoelectronic devices such as LEDs and solar cells. MOCVD is renowned for its repeatability and scalability, making it a preferred method for high-volume production of advanced materials.

3.3.4 Atomic Layer Deposition (ALD)

Mechanism and Benefits: ALD is a highly controlled, layer-by-layer deposition technique used to create ultra-thin films with atomic-level precision. The process involves alternating exposure of the substrate to different precursor gases, allowing each precursor to react only with the previous layer. This sequential reaction process ensures that only one atomic layer is deposited in each cycle, providing unparalleled control over film thickness and uniformity and its step by step process in an ALD reactor could be visualized through Fig. 3.3.

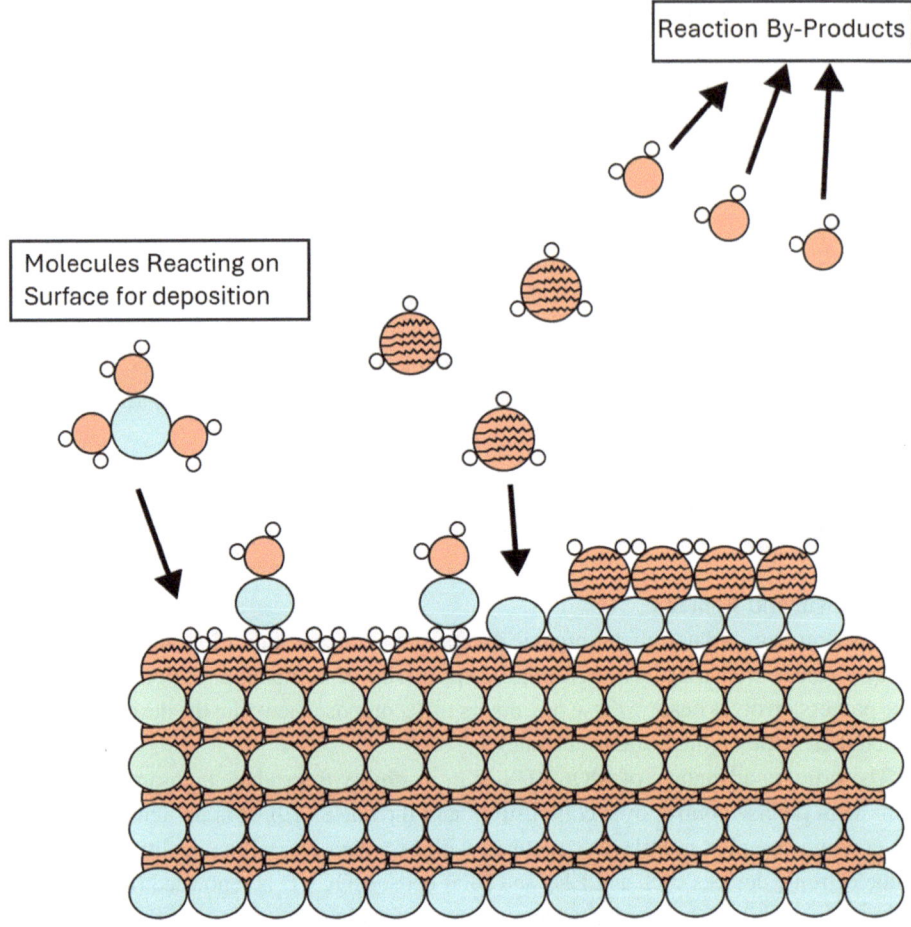

Fig. 3.2 Deposition Process in Metal-Organic Chemical Vapor Deposition (MOCVD)

The primary advantage of ALD is its ability to produce extremely conformal coatings, even on complex or high-aspect-ratio surfaces, which is essential in advanced semiconductor applications. ALD is widely used in applications requiring precise thickness control, such as gate dielectrics in transistors, insulating barriers, and protective coatings for batteries and optical devices. Its atomic precision makes ALD an invaluable technique for next-generation microelectronics, offering consistent, high-quality films with excellent repeatability across large-scale manufacturing.

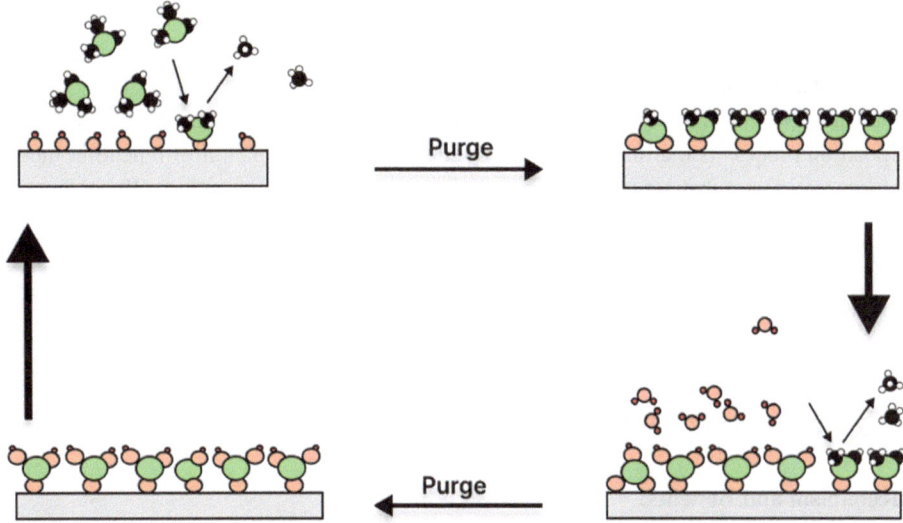

Fig. 3.3 Step-by-Step process in an atomic layer deposition (ALD) reactor

3.3.5 Sol-Gel CVD

Mechanism and Benefits: Sol-Gel CVD combines the sol-gel process with chemical vapor deposition to deposit oxide and composite films. In this method, a liquid precursor solution (or "sol") containing metal-organic compounds is applied to the substrate. This solution undergoes hydrolysis and polycondensation to form a gel-like network, which is then heat-treated to form a dense oxide layer on the substrate.

The main benefit of Sol-Gel CVD is its versatility and cost-effectiveness. It operates at relatively low temperatures compared to traditional CVD processes, making it suitable for temperature-sensitive substrates. Sol-Gel CVD allows for the deposition of a wide range of oxides and composite materials with excellent control over thickness and material properties. It is commonly used for optical coatings, dielectric layers, and protective films in applications such as sensors, thin-film transistors, and transparent conductive oxides. Its adaptability and simplicity make Sol-Gel CVD a valuable technique for research and specialized applications in electronics and materials science.

3.3.6   Hot-Wall CVD

Mechanism and Benefits:
Hot-Wall CVD heats the entire reaction chamber to a uniform temperature, ensuring that the deposition process is thermally stable throughout the space. This uniform heating method helps to avoid the substrate temperature variations typically found in cold-wall reactors [20].

One significant benefit of Hot-Wall CVD is the uniformity of the deposited films across large substrates, minimizing material wastage and enhancing the overall efficiency of material usage. This method is particularly useful for applications that require large-scale production with consistent properties across batches [21].

3.3.7   Cold-Wall CVD

Mechanism and Benefits:
Cold-Wall CVD focuses on heating only the substrate while keeping the chamber walls relatively cool. This configuration allows for rapid substrate heating and cooling, reducing thermal stresses and improving process turnaround times [22].

The primary advantage of Cold-Wall CVD is its rapid deposition rates and quick adaptation to changes in process parameters, which can be crucial for experimental designs and specialized applications requiring high flexibility. This method is especially valuable for materials that degrade or decompose under prolonged heat exposure.

3.4   CVD Materials in Semiconductor Packaging

CVD enables the deposition of a wide array of materials essential for the construction and performance of semiconductor devices. Each category of materials—dielectrics, conductors, and semiconductors—serves distinct functions within a device, from insulation and protection to conduction and active device operation.

3.4.1   Dielectrics

Silicon Dioxide (SiO_2)
Silicon dioxide, or silica, is one of the most commonly used dielectric materials in the semiconductor industry [23]. Deposited via CVD, (SiO_2) forms an essential insulating layer between conductive elements, preventing electrical shorts and crosstalk between signal lines, which is critical for maintaining the integrity and performance of the circuit.

Properties and Benefits:

Electrical Insulation: High dielectric strength prevents electron movement between conductive layers, essential for reliable device operation.

Thermal Stability: (SiO_2) maintains its properties under high thermal loads, making it suitable for devices experiencing wide temperature ranges.

Chemical Inertness: Resistant to most solvents, acids, and bases, protecting underlying structures from environmental damage.

Mechanism of Deposition: (SiO_2) is typically deposited using TEOS (Tetraethyl Orthosilicate) in either a plasma-enhanced or low-pressure CVD process, allowing for controlled growth of the oxide layer at temperatures suitable for various substrate materials [24].

Silicon Nitride (Si_3N_4) is another crucial dielectric material in semiconductor packaging, favored for its excellent mechanical and chemical properties.

Properties and Benefits:

Mechanical Durability: Provides robust protection against mechanical stresses and external impacts.

High Temperature Resistance: Maintains integrity at high temperatures, suitable for high-power applications.

Moisture and Chemical Resistance: Acts as an effective barrier against moisture and corrosive chemicals, enhancing device longevity.

Mechanism of Deposition: Silicon nitride is deposited through reactions involving silane (SiH_4) and ammonia (NH_3) gases in a LPCVD system, which facilitates the formation of a dense, uniform layer that adheres well to various substrates.

3.4.2 Conductors

Tungsten (W) is used in semiconductor devices primarily for its high melting point and excellent conductivity. It serves as a barrier and plug material in vias and contacts, connecting different layers of the device electrically.

Properties and Benefits:

High Conductivity: Efficiently transports electrons, minimizing energy loss and heat generation.

High Melting Point: Remains stable at temperatures where other metals might fail, essential for high-performance applications.

Mechanism of Deposition: Tungsten is deposited using (WF_6) (Tungsten Hexafluoride) as a precursor in a CVD process that involves hydrogen reduction at relatively low temperatures, allowing for controlled deposition on sensitive materials.

Copper (Cu) is increasingly used in semiconductor devices due to its superior electrical conductivity, particularly in interconnects where signal integrity is critical.

Properties and Benefits:

Superior Electrical Conductivity: Ensures minimal signal loss and lower power consumption.

Thermal Conductivity: Efficiently dissipates heat, helping to maintain stable device temperatures.

Mechanism of Deposition: Copper is typically deposited via CVD processes that involve organocopper precursors, which decompose on the heated substrate to form pure copper films.

3.4.3   Semiconductors

Gallium arsenide (GaAs) is a compound semiconductor used extensively in high-speed, high-frequency applications. It offers superior electron mobility compared to silicon, making it ideal for RF components and advanced computing [25].

Properties and Benefits:

High Electron Mobility: Allows for faster electron transport, enhancing device speed and efficiency.

Direct Bandgap: Enables efficient light emission, useful in optoelectronic devices like LEDs and laser diodes.

Mechanism of Deposition: GaAs is typically deposited using MOCVD, where precursors like trimethylgallium (TMG) and arsine (AsH_3) react at elevated temperatures to form crystalline layers on the substrate [26].

Gallium Nitride (GaN) is another vital semiconductor material, especially in power electronics and blue/UV light-emitting devices. It is known for its high thermal conductivity and robustness under harsh conditions [27].

Properties and Benefits:

High Thermal Conductivity: Manages heat effectively, essential for high-power applications.

High Breakdown Voltage: Supports higher voltages than silicon, crucial for power transistors.

Mechanism of Deposition: GaN is deposited using MOCVD, involving precursors like trimethylgallium and ammonia. The process allows for the formation of highly pure and crystalline GaN suitable for electronic and optoelectronic applications.

3.5   Applications of CVD in Packaging

3.5.1   CVD in TSVs

In the fabrication of Through-Silicon Vias, CVD plays a crucial role in depositing both the insulating and barrier layers necessary to maintain the structural and electrical integrity of the vias [28]. For insulating layers, materials like silicon dioxide or silicon nitride are deposited using LPCVD or PECVD techniques. These methods are preferred for their ability to achieve uniform, conformal coatings over the irregular shapes of the vias, ensuring complete coverage and preventing electrical leakage between the via and the surrounding substrate.

For barrier layers, materials such as titanium nitride (TiN) or tantalum (Ta) are deposited using ALD [29]. These barrier layers are essential not only for protecting the structural elements of the via but also for ensuring that the via remains conductively efficient and free from contaminants that could degrade its performance.

3.5.2   CVD in RDLs

The formation of Redistribution Layers heavily relies on CVD to deposit both dielectric isolation layers and the conductive metal traces. PECVD is typically employed to lay down dielectric materials such as silicon dioxide or low-k dielectrics. The process parameters are finely tuned to optimize the film's properties for electrical isolation while ensuring the film conforms to the underlying topography, which can be complex due to the presence of multiple layers and structures.

Metallic conductors such as copper are deposited using techniques like chemical vapor deposition that allow for the filling of finely patterned features typical of RDLs. These CVD processes are designed to achieve high-purity metal deposition, which is crucial for minimizing resistance and enhancing the electrical performance of the interconnects. The ability of CVD to deposit metals uniformly across the intricate patterns of RDLs ensures reliable electrical pathways that are critical for device functionality.

3.5.3   CVD in Protective Coatings

For protective coatings, materials such as silicon carbide (SiC) or diamond-like carbon (DLC) are typically deposited via PECVD or a specialized form of CVD that allows for the formation of amorphous carbon films [30]. These materials are selected for their exceptional hardness and resistance to environmental factors such as moisture and corrosive chemicals. The deposition process is controlled to produce coatings that are not only thick enough to provide effective protection but also possess the necessary adhesion and compactness to withstand physical impacts and abrasion.

The CVD techniques used for protective coatings are adjusted to balance the mechanical properties of the coatings with their functional requirements. For instance, the deposition conditions may be optimized to enhance the barrier properties against moisture ingress while maintaining sufficient flexibility to resist cracking under mechanical stress. This balance is critical for maintaining the longevity and reliability of semiconductor devices exposed to harsh operating conditions.

3.6   The Broad Reach of CVD: Enhancing NEMS, Photonics, and Encapsulation Technologies

3.6.1   CVD in MEMS and NEMS Fabrication

CVD plays a pivotal role in the fabrication of MEMS/NEMS by enabling the deposition of tailored materials with the required properties [31].

Silicon Carbide Films via CVD: Silicon carbide (SiC) is a widely used material in MEMS/NEMS devices due to its robust mechanical and thermal properties. Traditional CVD processes for SiC involve high temperatures that can compromise the integrity of underlying layers. Recent advancements have focused on refining these processes to enable lower temperature depositions. Developments in ALD, a subset of CVD, have been particularly promising, allowing for the creation of SiC films with enhanced uniformity and conformality at significantly reduced temperatures. This capability is crucial for producing MEMS/NEMS devices with more complex structures and improved durability.

Innovations and Developments: The ongoing development of CVD techniques, including the use of nanostructured materials, has improved the mechanical properties of MEMS/NEMS devices significantly, as evidenced by nanoindentation studies. These advancements not only enhance device performance but also extend their potential applications in areas requiring precise mechanical resilience.

3.6.2   CVD for Encapsulation Layer Deposition

Encapsulation layers are essential for protecting sensitive microelectronics from environmental exposure and mechanical damage, particularly in devices used in harsh or human-body environments. CVD processes, including LPCVD and PECVD, are instrumental in depositing high-integrity encapsulation materials [32].

Polymeric and Ceramic Encapsulants via CVD: Recent CVD innovations have expanded the range of materials suitable for encapsulation, including advanced polymers and ceramics. PECVD, for instance, has been adapted to deposit complex polymeric compounds that provide superior environmental sealing and mechanical protection. These advancements are crucial for applications such as implanted medical devices, where material compatibility with the biological environment is paramount [33].

Challenges and Advancements: The encapsulation materials developed through CVD must withstand diverse environmental stresses, including temperature fluctuations and chemical exposure. The ongoing refinement of CVD processes aims to enhance the durability and longevity of these materials, thus improving the reliability of the encapsulated devices.

3.6.3   CVD in Silicon Photonics

Silicon photonics is an area where CVD's ability to deposit precise, uniform layers becomes invaluable [34]. This technology integrates optical functionalities directly onto silicon chips, facilitating high-speed data transmission capabilities that are crucial for next-generation computing and telecommunication systems.

Building Blocks of Silicon Photonics via CVD: In silicon photonics, CVD is used to deposit both the active and passive layers required for the optical components, such as waveguides [35], modulators [36], and detectors [37]. The ability to control the thickness and composition of these layers with high precision ensures optimal optical performance and integration compatibility with existing silicon-based electronics.

Versatility and Adaptation of CVD: The versatility of CVD in depositing various doped layers and dielectrics allows for the engineering of structures with tailored refractive indices, essential for effective light guidance and confinement. Furthermore, the low-temperature processing capabilities of advanced CVD techniques help minimize thermal stresses that could potentially alter the optical properties of the materials.

3.7   CVD-Based Solutions for Critical Vulnerabilities

3.7.1   Thermal Stress

Vulnerability: As devices become smaller and more powerful, the amount of heat they generate increases. This heat must be efficiently managed to prevent degradation of the device performance and reliability. Poor thermal management can lead to overheating, which can accelerate material degradation, affect device performance, and reduce the overall device lifespan.

CVD-Based Mitigation: CVD can be used to deposit high thermal conductivity materials, such as diamond-like carbon or silicon carbide, as heat spreaders or thermal interface materials. These materials can effectively dissipate heat away from hot spots, thereby enhancing device reliability and performance.

3.7.2   Chemical and Moisture Ingress

Vulnerability: Semiconductor devices are sensitive to environmental factors, including moisture and corrosive chemicals, which can penetrate the device and cause corrosion or short-circuiting. This is particularly problematic in harsh environments and can drastically reduce the device's operational life.

CVD-Based Mitigation: Utilizing CVD to apply robust protective coatings such as silicon nitride or silicon carbide can provide a high-quality, impermeable barrier to moisture and chemicals. These coatings are highly effective at sealing the device from environmental exposure, thereby ensuring the longevity and reliability of the components.

3.7.3   Mechanical Stress and Fracture

Vulnerability: Mechanical stresses from shock, vibration, or thermal expansion mismatches can lead to fractures or delamination within the semiconductor package. Such physical damage not only impacts the current functionality but also predisposes the device to further damage and failure.

CVD-Based Mitigation: CVD processes can deposit hard, wear-resistant, and flexible coatings that can provide a mechanical reinforcement layer to the semiconductor devices. Materials like diamond-like coatings or composite structures designed via CVD can absorb and redistribute stresses, reducing the risk of fracture and enhancing the mechanical robustness of the package.

3.7.4   Electromigration

Vulnerability: In high-density packages, the increased current density can lead to electromigration, where metal atoms are displaced by the momentum of the conducting electrons, leading to voids and eventually circuit failures. This issue is compounded as the scale of interconnections shrinks in modern devices.

CVD-Based Mitigation: Applying robust barrier layers using ALD, a subtype of CVD, can significantly improve the resistance of metal interconnects to electromigration. Materials like tantalum nitride or titanium nitride can be precisely deposited to form effective barriers that prevent atom displacement and maintain the integrity of the interconnects.

3.7.5   Electrical Leakage and Crosstalk

Vulnerability: As the density of components and their proximity increase, electrical leakage and signal crosstalk become significant issues, degrading the performance of the device and causing functional errors.

CVD-Based Mitigation: High-quality dielectric materials such as low-k dielectrics can be deposited using CVD techniques to provide effective electrical insulation. This reduces leakage currents and minimizes crosstalk between closely packed lines, thereby ensuring the fidelity of the signals within the device.

3.7.6   Contamination During Fabrication

Vulnerability: Contamination by particles or unwanted chemical residues during the packaging process can lead to defects that compromise device performance.

CVD-Based Mitigation: The precision and controlled environment of CVD processes ensure high purity and cleanliness during the deposition of materials, reducing the risk of contamination. Additionally, the conformal nature of CVD films ensures complete coverage, even on complex topographies, which helps seal off the underlying layers from potential contaminants introduced in later processing steps.

3.8   Future Trends and Innovations in CVD Technology

As the demands on semiconductor devices continue to increase with advancements in technology, the development of CVD methodologies keeps pace, evolving to meet the emerging challenges of device miniaturization and complexity. Below, we delve deeper into two significant areas of innovation within CVD technology: the creation of nanostructured materials and the progression towards more sustainable, eco-friendly processes.

3.8.1   Nanostructured Materials

Advancements in Nanostructured CVD:
Recent developments in CVD technology have enabled the controlled synthesis of nanostructured materials, which are engineered at the molecular level to achieve superior performance characteristics. Nanostructures, including nanotubes, nanowires, and quantum dots, can be precisely deposited using specialized CVD techniques. These structures often exhibit unique electrical, thermal, and mechanical properties not found in bulk materials, primarily due to their high surface area to volume ratios and quantum mechanical effects at very small dimensions.

Properties and Applications:

Electrical Properties: Nanostructured materials such as carbon nanotubes (CNTs) and graphene have shown exceptional electrical conductivity and electron mobility, making them ideal for use in next-generation electronic components, including transistors, sensors, and conductive films.

Thermal Properties: Nanostructured materials can also significantly improve thermal management in devices. For example, diamond-like carbon coatings have been developed via CVD to enhance heat dissipation capabilities in high-performance processors.

Mechanical Properties: Enhanced mechanical properties such as strength and flexibility are critical, especially in wearable electronics and flexible displays. Nanostructured materials deposited via CVD can provide the necessary durability and flexibility without compromising device performance.

Innovative CVD Techniques for Nanostructures:

Innovations in CVD technology have focused on improving the deposition conditions to grow uniform nanostructured materials with controlled orientation and alignment, critical for integrating these materials into existing semiconductor manufacturing processes. Techniques like PECVD have been adapted to facilitate the growth of vertically aligned nanotubes, which are essential for applications requiring high-density, high-efficiency thermal and electrical conductors.

3.8.2   Eco-Friendly CVD Processes Environmental Impact Reduction

The semiconductor manufacturing industry faces increasing pressure to reduce its environmental footprint, prompting innovations in CVD processes to become more sustainable. Efforts are concentrated on developing CVD methods that lower energy consumption and minimize the use of hazardous chemicals.

Sustainable Practices in CVD:

Reduced Energy Consumption: Advanced CVD systems are being designed to operate at lower temperatures and pressures, significantly reducing the energy required for deposition. Innovations such as ALD have shown that it is possible to achieve high-quality depositions at reduced thermal budgets, decreasing overall energy usage.

Use of Green Chemical Precursors: There is a growing focus on replacing toxic and hazardous chemical precursors with more environmentally friendly alternatives. Research is ongoing into precursors that yield fewer byproducts and can be decomposed into harmless substances, thus minimizing chemical waste and potential environmental contamination.

Impact and Benefits:

The shift towards eco-friendly CVD processes not only aligns with global sustainability goals but also reduces operational costs for semiconductor manufacturers. By investing in greener technologies, the industry can mitigate its environmental impact while maintaining the high standards required for advanced semiconductor production.

3.9   Strategic Insights Into the CVD Technology Ecosystem

The CVD market is distinguished by a few influential players whose innovations set the pace for industry standards and technological advancements. Companies like Applied Materials, Tokyo Electron Limited, and ASM International are not just manufacturers but pioneers shaping the future of semiconductor technologies.

Applied Materials leads with its AKT-PECVD systems, known for their precision in depositing complex multilayer structures essential for high-definition LCD displays. This technology underscores Applied Materials' focus on enhancing throughput while reducing material waste and energy consumption, aligning with the industry's push towards sustainability.

Tokyo Electron Limited complements these advancements with its tailored CVD systems that cater to specific needs across various semiconductor processes, emphasizing flexibility and performance optimization.

3.9.1   Collaborative Innovations and Sector Growth

These industry leaders thrive on collaboration that spans across the value chain, encompassing material suppliers, equipment manufacturers, and end users. Such collaborations are crucial for integrating novel technologies into established production lines and for co-developing products that meet emerging market needs.

Strategic Partnerships: For instance, collaborations between companies like ASM International and research institutions have led to the development of advanced PECVD tools that drastically reduce deposition times and energy use, thereby enhancing the economic and environmental aspects of semiconductor manufacturing.

Supply Chain Synergies: Close interactions with material suppliers ensure that CVD equipment manufacturers have access to the latest high-purity precursors needed for developing next-generation semiconductor devices, thereby maintaining a competitive edge in the market.

3.9.2   Market Dynamics and Future Prospects

The CVD market's expansion is driven by the growing demand for more sophisticated electronic devices, which require complex and reliable semiconductor components. As the industry continues to innovate, CVD technology vendors are well-positioned to capitalize on emerging opportunities.

Market Expansion: The global expansion of the CVD market is facilitated by increasing investments in areas like solar energy and electronics, where CVD techniques are vital for producing high-performance, cost-effective solutions.

Technological Evolution: Continuous research into low-temperature CVD processes and green manufacturing practices highlights the sector's commitment to innovation and sustainability. These efforts are crucial for minimizing the environmental impact of production processes and for meeting stricter global regulations.

Challenges and Strategic Responses Despite the sector's growth, challenges such as high operational costs, complex technology integration, and stringent environmental regulations pose significant hurdles. Strategic responses from CVD technology providers include investing in R&D to overcome these challenges and implementing more efficient, less resource-intensive manufacturing processes.

Innovative Solutions: Developments like high-density plasma-enhanced CVD (HDPECVD) systems from companies like Plasma-Thermo LLC illustrate how adapting CVD technologies to produce thinner, more uniform films can address issues related to device miniaturization and performance optimization.

> *Conclusion*:
> In this chapter, we explored the vital role of CVD technology in semiconductor manufacturing. CVD's precision in layering materials is crucial for advanced applications like MEMS, NEMS, and silicon photonics, addressing challenges of miniaturization and functionality. Innovations in eco-friendly and nanostructured materials highlight CVD's adaptability to evolving industry demands. With ongoing advancements and strategic industry collaborations, CVD is poised to further revolutionize semiconductor technologies, ensuring its pivotal role in shaping the future landscape of electronics and beyond.

References

1. Walke, Santosh, et al. "A review on copper chemical vapour deposition." *Materials Today: Proceedings* (2023).
2. Schmit, Claire E., and Gregory S. Girolami. "Volatile N, N-dialkyl–dialdiminate complexes of magnesium and zinc as possible chemical vapor deposition precursors." *Polyhedron* 256 (2024): 116985.
3. Li, Qizhong, et al. "Improvement of SiC deposition uniformity in CVD reactor by showerhead with baffle." *Journal of Crystal Growth* 615 (2023): 127255.

4. Gupta, Tejendra K., et al. "Chemical vapor deposition of ferrite thin films." In *Ferrite Nanostructured Magnetic Materials*. Woodhead Publishing, 2023, pp. 293–308.

5. Shi, Haolian, et al. "Terahertz Nondestructive Characterization of Conformal Coatings for Microelectronics Packaging." *IEEE Transactions on Components, Packaging and Manufacturing Technology* (2024).

6. Huang, James. *Selective Oxide Pulsed Chemical Vapor Deposition for Dielectric on Metal and Dielectric on Dielectric*. University of California, San Diego, 2023.

7. Lausecker, Clément, David Muñoz-Rojas, and Matthieu Weber. "Atomic layer deposition (ALD) of palladium: from processes to applications." *Critical Reviews in Solid State and Materials Sciences* 49.5 (2024): 908–930.

8. Patel, Amit Kumar, Ashish Jyoti Borah, and Anchal Srivastava. "Optimized APCVD method for synthesis of monolayer H-Phase VS2 crystals." *Oxford Open Materials Science* 3.1 (2023): itad020.

9. Zhou, Jicheng, et al. "Simulation and optimization of polysilicon thin film deposition in a 3000 mm tubular LPCVD reactor." *Solar Energy* 253 (2023): 462–471.

10. Bertran-Serra, Enric, et al. "Advancements in plasma-enhanced chemical vapor deposition for producing vertical graphene nanowalls." *Nanomaterials* 13.18 (2023): 2533.

11. Bloem, J. "High chemical vapour deposition rates of epitaxial silicon layers." *Journal of Crystal Growth* 18.1 (1973): 70–76.

12. Nikhar, Tanvi, and Sergey V. Baryshev. "Evidence of gas phase nucleation of nanodiamond in microwave plasma assisted chemical vapor deposition." *AIP Advances* 14.4 (2024).

13. Chowdhury, Mohammad A., Dewan M. Nuruzzaman, and Mohammad L. Rahaman. "The effect of gas flow rate on the thin film deposition rate on carbon steel using thermal CVD." *International Journal of Chemical Reactor Engineering* 9.1 (2011).

14. Tang, Haonan, et al. "Nucleation and coalescence of tungsten disulfide layers grown by metalorganic chemical vapor deposition." *Journal of Crystal Growth* 608 (2023): 127111.

15. Lu, Chen-Hsuan, Duxing Hao, and Nai-Chang Yeh. "A perspective of recent advances in PECVD-grown graphene thin films for scientific research and technological applications." *Materials Chemistry and Physics* (2024): 129318.

16. Perez, Christopher, et al. "High thermal conductivity of submicrometer aluminum nitride thin films sputter-deposited at low temperature." *ACS Nano* 17.21 (2023): 21240–21250.

17. Makarenko, Alexander M., Dzmitry H. Zaitsau, and Kseniya V. Zherikova. "Metal-organic chemical vapor deposition precursors: Diagnostic check for volatilization thermodynamics of scandium (III) -diketonates." *Coatings* 13.3 (2023): 535.

18. Nong, Mingtao, et al. "Epitaxial AlN film with improved quality on Si (111) substrates realized by boron pretreatment via MOCVD." *Applied Physics Letters* 124.17 (2024).

19. Tian, Jing, et al. "Simulation of Epitaxial Growth of Silicon Carbide in a Horizontal Hot-wall CVD Reaction Chamber." In *2023 20th China International Forum on Solid State Lighting & 9th International Forum on Wide Bandgap Semiconductors (SSLCHINA: IFWS)*. IEEE, 2023.

20. Gogova, D., et al. "High crystalline quality homoepitaxial Si-doped -Ga2O3 (010) layers with reduced structural anisotropy grown by hot-wall MOCVD." *Journal of Vacuum Science & Technology A* 42.2 (2024).

21. Anderson, Andrew. *Development of Cold-Wall Chemical Vapour Deposition Systems for the Growth of Carbon Nanotubes and Carbon Fibres*. 2023.

22. Fujino, K., et al. "Silicon dioxide deposition by atmospheric pressure and low-temperature CVD using TEOS and ozone." *Journal of The Electrochemical Society* 137.9 (1990): 2883.

23. Yan, Junjie, et al. "Microstructure and oxidation behavior of SiOC coatings on C/C composites co-deposited with HMDS and TEOS by using CVD process." *Ceramics International* 50.5 (2024): 7888–7896.

24. Rocha-Arredondo, L. E., et al. "Raman study of directly synthetized graphene oxide films on Si, SiO2/Si and GaAs by remote-catalyzed CVD." *Physica B: Condensed Matter* 669 (2023): 415302.
25. Kim, Hankyu, et al. "Adsorption mechanism of dimeric Ga precursors in metalorganic chemical vapor deposition of gallium nitride." *Journal of Vacuum Science & Technology A* 41.6 (2023).
26. Wang, Dong, et al. "Theoretical adjustment of metalorganic chemical vapor deposition process parameters for high-quality gallium nitride epitaxial films." *Physics of Fluids* 35.3 (2023).
27. Jousseaume, Vincent, et al. "Wafer Scale Insulation of High Aspect Ratio Through-Silicon Vias by iCVD." *ACS Applied Materials & Interfaces* (2024).
28. Lee, Baek-Ju, et al. "The Effect of deposition temperature of TiN thin film deposition using thermal atomic layer deposition." *Coatings* 13.1 (2023): 104.
29. Malisz, Klaudia, Beata Świeczko-Żurek, and Alina Sionkowska. "Preparation and Characterization of Diamond-like Carbon Coatings for Biomedical Applications—A Review." *Materials* 16.9 (2023): 3420.
30. Arjmandi-Tash, Hadi, et al. "Large scale integration of CVD-graphene based NEMS with narrow distribution of resonance parameters." *2D Materials* 4.2 (2017): 025023.
31. Zhang, Buyue, et al. "Recent Achievements for Flexible Encapsulation Films Based on Atomic/Molecular Layer Dep Deposition." *Micromachines* 15.4 (2024): 478.
32. Chou, Fang-Yu, et al. "Vapor-Deposited Polymer Films and Structure: Methods and Applications." *Organic Materials* 5.02 (2023): 118–138.
33. Taverne, Mike PC, et al. "Conformal CVD-Grown MoS2 on Three-Dimensional Woodpile Photonic Crystals for Photonic Bandgap Engineering." *ACS Applied Optical Materials* 1.5 (2023): 990–996.
34. Fang, Yuxi, et al. "Recent progress of supercontinuum generation in nanophotonic waveguides." *Laser & Photonics Reviews* 17.1 (2023): 2200205.
35. Yang, Yiqian, Andrew Forbes, and Liangcai Cao. "A review of liquid crystal spatial light modulators: devices and applications." *Opto-Electronic Science* 2.8 (2023): 230026-1.
36. Grönberg, Fredrik, et al. "The effects of intra-detector Compton scatter on low-frequency DQE for photon-counting CT using edge-on-irradiated silicon detectors." *Medical Physics* 51.7 (2024): 4948–4969.
37. Vir Singh, Man, Ajay Kumar Tiwari, and Rajeev Gupta. "Catalytic chemical vapor deposition methodology for carbon nanotubes synthesis." *ChemistrySelect* 8.32 (2023): e202204715.

Etching Techniques and Applications in Advanced IC Packaging

4

4.1 Introduction

Etching is a vital process in microelectronics packaging, used to selectively remove material from a surface to create patterns, channels, and cavities that define a device's physical and electrical structure. This process enables the precise shaping of substrates and layers, facilitating connections between different materials and components. Without controlled etching, achieving the intricate designs needed for advanced electronic packaging—like RDLs, vias, and interconnects—would be impossible [1].

In the context of packaging, etching is essential for enabling compact, high-performance devices. With microelectronics packaging moving towards higher density integration and multi-layer architectures, etching allows engineers to create the necessary features for electrical interconnections within a small footprint. As packaging shifts towards three-dimensional designs the role of etching becomes even more critical, ensuring that every layer aligns perfectly without disrupting signal integrity.

Etching as illustrated in Fig. 4.1. Typically follows deposition steps like CVD (covered in the previous chapter), where thin layers of material are added to a substrate. After these layers are patterned through photolithography [2], etching removes the unprotected areas to define the desired structure. A successful etch process ensures precise depth and shape, which is critical to avoid device failure or performance degradation. The challenge lies in achieving this precision while maintaining a high throughput to meet industrial production demands.

Two primary types of etching—wet etching [3] and dry etching [4]—offer distinct advantages. Wet etching uses liquid chemicals to dissolve materials selectively, while dry etching involves gas-phase reactions and plasma bombardment. Packaging engineers often employ both methods, balancing simplicity, precision, and scalability based on the material and

© The Author(s), under exclusive license to Springer Nature Switzerland AG 2025
N. Asadizanjani et al., *Introduction to Microelectronics Advanced Packaging Assurance*,
Synthesis Lectures on Engineering, Science, and Technology,
https://doi.org/10.1007/978-3-031-86102-4_4

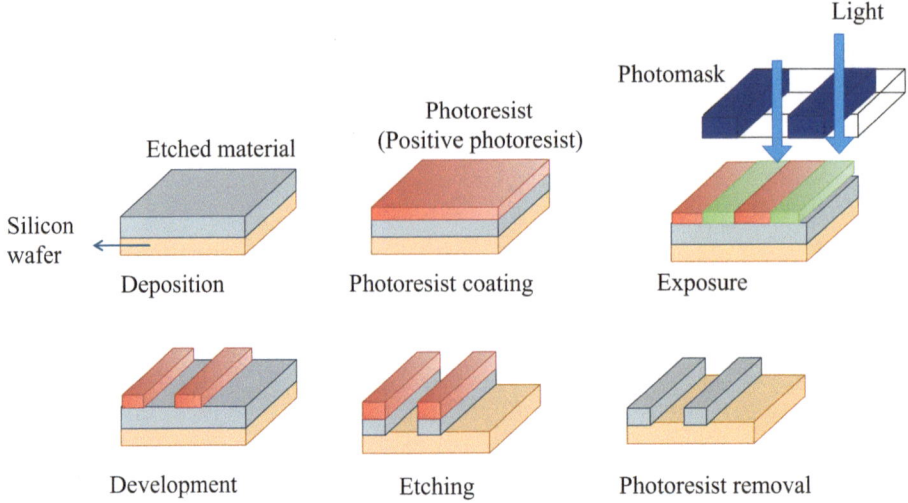

Fig. 4.1 Overview of the etching process

design needs. For instance, wet etching may be preferred for bulk material removal, while dry etching ensures the sharp, vertical profiles needed for modern packaging interconnects. Each technique must be carefully selected and optimized to align with the process requirements and avoid defects like undercutting or over-etching [5].

In today's microelectronics landscape, etching is no longer just a fabrication step—it is part of the design strategy, influencing decisions about material selection, process flows, and device architecture. With increasing focus on smaller nodes and tighter geometries, innovations in etching are becoming essential. Techniques like atomic layer etching (ALE) promise even greater precision, removing material layer-by-layer to meet the challenges posed by next-generation packaging technologies [6].

As the industry moves forward, etching will play a central role in enabling higher integration, better thermal management, and enhanced reliability. Mastering both traditional and emerging etch processes is essential for engineers striving to design packages that are not only functional but also robust, ensuring that devices can withstand the stresses of operation in diverse environments.

4.2   Classification of Etching Techniques

Etching is classified into two primary categories: wet etching and dry etching. Each technique offers unique advantages and challenges, making them suitable for specific processes based on the material, feature size, and design requirements. Understanding the differences between these methods is crucial for selecting the optimal etching approach (Fig. 4.2).

4.2.1 Wet Etching

Wet etching is a chemical process that involves immersing or exposing the target material to a liquid etchant [7]. The etchant dissolves the exposed material through chemical reactions, resulting in controlled removal. Wet etching is generally used for processes where precise feature control is not critical but large-scale or bulk material removal is required.

4.2.2 Mechanism of Wet Etching

Diffusion of the Etchant: The liquid etchant penetrates the surface through exposed areas and reaches the material to be etched.

Chemical Reaction: The etchant reacts with the substrate material, breaking chemical bonds and dissolving the material into the solution.

Byproduct Removal: The dissolved material diffuses away from the surface, making room for further chemical interactions.

Wet etching is usually isotropic [8], meaning the material is removed uniformly in all directions. This isotropy can lead to undercutting, where the etchant removes material beneath the masking layer, resulting in rounded or irregular sidewalls [9].

4.2.3 Advantages of Wet Etching

Simple and Cost-effective: It requires basic equipment and can be easily implemented with minimal complexity.

High Selectivity: Wet etchants can be highly selective, removing one material without affecting others, making it useful for multi-layer structures.

High Etch Rates: Wet etching offers fast processing speeds, making it suitable for applications that require the removal of thick layers or bulk material.

4.2.4 Challenges of Wet Etching

Limited Anisotropy: The isotropic nature of wet etching can lead to lateral etching, compromising pattern accuracy and limiting its use in high-resolution applications.

Mask Compatibility Issues: Some chemical etchants may attack or degrade the masking material, reducing the process's reliability.

Chemical Waste Management: Etchants, such as hydrofluoric acid (HF), are hazardous and require careful handling and disposal protocols to ensure environmental and workplace safety [10].

4.2.5 Common Applications of Wet Etching

Oxide Removal: HF is commonly used to remove silicon dioxide (SiO_2) layers, crucial for preparing substrates for further processing.
Metal Etching: Phosphoric acid is used to etch aluminum, while ammonium persulfate can etch copper, both of which are essential for interconnection patterning [11].
Polymer Patterning: Wet etching is also applied to shape polymers such as polyimides, which are often used for mechanical support or dielectric insulation [12].

4.3 Dry Etching

Dry etching is a plasma-based process where material is removed through the interaction of reactive gases or ions with the substrate. It provides high precision and is indispensable for processes requiring anisotropic control [13], enabling the formation of sharp, vertical profiles. Dry etching operates in a vacuum environment, making it more complex but also more versatile than wet etching.

4.3.1 Mechanism of Dry Etching

Ion Bombardment: Plasma-generated ions bombard the surface of the material, physically sputtering it away through kinetic energy transfer [14].
Chemical Reactions: Reactive gas molecules interact with the material, forming volatile byproducts that can be pumped out of the vacuum chamber.
Energy Transfer: The plasma provides the necessary energy to sustain the chemical reactions and drive the removal of material.
This combination of physical sputtering and chemical etching ensures high anisotropy, making dry etching the preferred technique for creating precise vertical structures.

4.3.2   Types of Dry Etching

Reactive Ion Etching (RIE) RIE uses plasma to generate reactive ions that etch the material through a combination of chemical reactions and physical ion bombardment. RIE achieves anisotropic etching, producing well-defined sidewalls, making it suitable for applications requiring sharp profiles [15].

Examples: Chlorine-based plasmas are used for etching metals like aluminum. Fluorocarbon plasmas are ideal for etching SiO_2 and other dielectrics [16].

Deep Reactive Ion Etching (DRIE) DRIE is an advanced form of RIE designed to etch deep, high-aspect-ratio structures. This technique alternates between etching the material and depositing a passivation layer to protect the sidewalls from further etching [17].

Plasma Etching Plasma etching relies on neutral gas molecules, rather than ions, to chemically react with the substrate material. While it offers high selectivity, it lacks the physical component required for anisotropic control [18].

4.3.3   Advantages of Dry Etching

Superior Anisotropy: Dry etching allows for the creation of precise vertical profiles, crucial for producing nanoscale features.

Controlled Precision: The combination of physical and chemical processes ensures fine control over the etching depth and pattern.

Material Versatility: It can etch a wide variety of materials, including metals, dielectrics, and semiconductors.

4.3.4   Challenges of Dry Etching

Complex Equipment: Dry etching requires advanced equipment, such as vacuum systems and plasma generators, which makes the setup expensive and harder to maintain.

Slow Etch Rates: Due to the precision it offers, dry etching is generally slower than wet etching, making it unsuitable for bulk material removal.

Charging Effects: In some cases, ions can accumulate on dielectric surfaces, leading to localized electrical charging and potential damage to the material [19].

Aspect	Dry Etching	Wet Etching
Process Medium	Plasma or gas phase	Liquid chemical solution
Etch Directionality	Generally anisotropic (directional)	Typically isotropic (non-directional)
Precision	High precision, suitable for fine features	Lower precision, may lead to undercutting
Application	Used for high-resolution, complex micro- and nanostructures	Suitable for larger features or less critical patterning
Environmental Impact	May produce harmful gases, requires gas management systems	Involves chemical disposal, may have environmental concerns

Fig. 4.2 Dry etching versus wet etching: key differences

4.4   Process Parameters for High Precision Etching

Achieving high precision in etching processes is essential for creating the intricate structures required in modern microelectronics packaging. Several key parameters—selectivity, etch rate, and anisotropy—must be carefully controlled to ensure accuracy, avoid defects, and maintain the integrity of the design throughout the etching process. Each of these parameters plays a distinct role in balancing throughput, precision, and performance in both wet and dry etching techniques.

4.4.1   Etch Selectivity

Selectivity refers to the ratio of the etch rate of the target material to the etch rate of other exposed layers, such as the mask material or underlying films. High selectivity ensures that only the intended material is removed, leaving the other layers unaffected or minimally etched. Optimizing selectivity is critical because it prevents unwanted damage to the mask or underlying layers, ensuring the fidelity of the patterned design. This parameter becomes especially important when working with thin or sensitive materials, where even slight over-etching can compromise functionality.

For example, in RIE, high selectivity toward the photoresist mask ensures that the pattern transfers accurately onto the underlying material. If the etch process lacks sufficient selectivity, the mask may erode before the etching is complete, leading to feature distortion or defects. Similarly, during the etching of dielectrics like (SiO_2), the selectivity toward underlying silicon must be high enough to prevent damage to the substrate. Engineers optimize selectivity by adjusting the etch chemistry, plasma parameters, and masking materials to achieve the desired result without compromising other layers in the process stack.

4.4.2 Etch Rate

The etch rate measures how quickly material is removed from the substrate, typically expressed in nanometers or microns per minute. The etch rate is a crucial parameter because it directly impacts process throughput and efficiency. However, achieving the right balance between etch rate and precision can be challenging, as higher etch rates often come at the expense of fine control and accuracy.

In high-throughput manufacturing environments, faster etch rates are desirable to reduce production time. However, increased etch rates can lead to issues such as non-uniform removal, rough surfaces, or difficulty in stopping the process precisely at the desired depth. This trade-off becomes particularly important in processes like deep trench etching or TSV formation, where maintaining control over the depth and profile of the etched feature is critical. Engineers optimize the etch rate by carefully tuning process variables such as plasma power, gas flow rates, and pressure. Achieving the right etch rate ensures that the process meets both speed and precision requirements without compromising quality.

The general formula for calculating the **etch rate** (*ER*) is given by:

$$\text{Etch Rate (ER)} = \frac{\Delta h}{\Delta t} \tag{4.1}$$

where:

- Δh = Change in material thickness (or depth of material etched away)
- Δt = Time over which the material is etched

This formula is applicable to both **wet** and **dry etching** methods.

- In **wet etching**, the etch rate depends on factors such as chemical concentration, solution temperature, and agitation.
- In **dry etching**, the etch rate is influenced by plasma power, ion energy, gas composition, and pressure.

While this formula provides a basic etch rate calculation, more complex models may include variables to account for *directionality*.

4.4.3 Anisotropy

Anisotropy as depicted in Fig. 4.3 refers to the directionality of the etching process. An anisotropic etch removes material primarily in the vertical direction, creating sharp, well-defined sidewalls that are essential for critical microelectronic features such as vias, trenches, and RDLs. In contrast, isotropic etching removes material uniformly in all directions, which can lead to undercutting where the material beneath the mask is also etched resulting in rounded or imprecise features.

Anisotropic control is particularly important in advanced packaging, where high-aspect-ratio structures are needed. For example, in TSV fabrication, the etch process must create deep, narrow channels with vertical walls to enable electrical connections between stacked chips. Dry etching techniques, such as RIE and DRIE, offer superior anisotropy compared to wet etching, as the physical bombardment by plasma ions ensures that material is removed only from the exposed areas, minimizing lateral etching. In DRIE, alternating cycles of etching and sidewall passivation further enhance anisotropic control, allowing for deep trench formation with nearly vertical profiles.

Achieving optimal anisotropy requires careful adjustment of process parameters, such as ion energy, gas composition, and pressure, to ensure that the etch remains highly directional. Wet etching, on the other hand, tends to be more isotropic because the chemical solution dissolves the material in all directions, limiting its application in processes that demand tight geometric control. In packaging applications where fine interconnects or trenches are needed, anisotropic etching is indispensable to maintain alignment and avoid signal degradation caused by misaligned or distorted features.

4.4.4 Isotropy

Isotropy refers to the uniformity of the etching process in all directions. An isotropic etch removes material evenly from all exposed surfaces, resulting in rounded features and less precise sidewall control. This uniform etching can be advantageous in applications where a smooth, gradual removal of material is preferred or where critical feature definition is not required.

In isotropic etching, both vertical and lateral etching rates are similar, which can lead to undercutting beneath the mask. This undercutting can limit the precision of fine structures, making isotropic etching less suitable for applications that require high-aspect-ratio features, such as vias and trenches in advanced microelectronics.

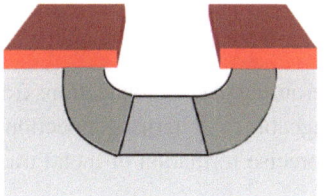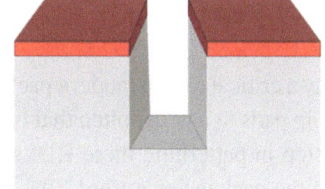

Fig. 4.3 **a** Isotropic etching versus **b** Anisotropic etching

However, isotropic etching as depicted in Fig. 4.3 has practical applications in scenarios where gradual material thinning or contouring is desired, such as in MEMS devices, where specific shapes and curves can benefit functionality. Wet etching techniques, including chemical baths, tend to exhibit isotropic behavior, as the etching solution dissolves the material in all directions.

While isotropy can be challenging for precision microelectronic applications, it remains valuable in processes that are less dependent on stringent dimensional accuracy. For example, in certain sensor or micromachining applications, isotropic etching helps shape components with rounded profiles. Additionally, isotropic wet etching is more cost-effective and easier to implement than highly controlled anisotropic techniques, making it a go-to choice in applications where vertical sidewalls are not a strict requirement.

Optimal isotropy can be achieved by carefully controlling factors such as temperature, chemical concentration, and etchant type to ensure uniform material removal. However, in applications demanding critical feature definition, such as high-density interconnects or narrow trenches, isotropic etching is limited by its inability to maintain sharp, vertical profiles essential for precise alignment and functionality.

4.5 Etching Across Packaging Processes: From RDLs to TSVs

Etching plays an essential role in the intricate processes involved in microelectronics packaging. It enables precise material removal, making it possible to define patterns, create interconnects, and prepare surfaces for bonding or encapsulation. With the continuous evolution toward higher integration, reduced form factors, and complex three-dimensional designs, etching—whether wet or dry—has become a crucial step in ensuring the functionality and reliability of modern packaging technologies. This chapter explores the various applications of etching in packaging, highlighting its role in RDL, TSVs, solder bump preparation, polyimide patterning, passivation, and MEMS packaging.

4.5.1 RDL Patterning

RDLs play a critical role in modern packaging by rerouting the I/O connections from closely spaced chip pads to a wider pitch that is more manageable for external connections. Etching is a key step in patterning these RDLs, allowing precise formation of metal traces on the surface of the package or wafer [20].

Wet Etching of Metal Layers: RDLs typically employ metal layers such as Cu or Al. Wet etching is a cost-effective approach for patterning these metals. For example, ammonium persulfate is often used to etch copper, while phosphoric acid works well for aluminum. Wet etching is particularly beneficial for large-scale RDL production, as it offers high throughput and relatively simple implementation. However, it requires careful process control to prevent undercutting, which can affect circuit integrity.

Dry Etching for Precise Patterning: Advanced packaging designs demand high precision, making dry etching essential for forming fine features with vertical sidewalls. RIE is often employed to etch oxide layers used as dielectric barriers in RDLs. The superior anisotropy of dry etching ensures accurate pattern transfer, minimizing lateral spread and ensuring performance at high densities.

4.5.2 Through-Silicon Vias (TSVs) Formation

Deep Reactive Ion Etching (DRIE): The formation of TSVs relies heavily on DRIE, a specialized etching technique that creates high-aspect-ratio structures. DRIE operates in cyclic steps, alternating between etching the silicon and depositing a passivation layer to protect the sidewalls from further etching. This ensures the precise verticality required for reliable electrical connections. The ability to create deep, narrow trenches without compromising accuracy makes DRIE indispensable for TSV formation.

4.5.3 Solder Bump Formation and Pad Preparation

Oxide Removal with Wet Etching: Before depositing solder bumps, it is necessary to remove native silicon dioxide (SiO_2) layers from the contact pads. Hydrofluoric acid (HF) is commonly used for this purpose, as it ensures the surface is clean and free from insulating oxides, improving adhesion and electrical conductivity. This step is vital in ensuring the quality and reliability of flip-chip bonds.

Plasma Etching for Residue Cleaning: After solder bumping, plasma etching is employed to remove organic residues or polymer films left from prior processes. This cleaning step ensures a contamination-free surface, reducing the risk of weak or defective solder joints.

4.5.4   Polyimide Patterning for Flexible Packaging

Polyimides are frequently used in packaging as insulating layers and flexible substrates. They provide mechanical support while maintaining thermal stability and electrical insulation, making them ideal for flexible electronics.

Wet Etching for Polyimide Films: Wet etching is an effective method for patterning polyimide films, enabling the formation of openings or circuit paths. Potassium hydroxide (KOH) or similar solutions are often used to shape these films. The etching process must be carefully controlled to ensure uniformity and prevent damage to other packaging layers.

Applications in Flexible Electronics: Polyimide patterning is essential in the development of flexible printed circuit boards (FPCBs), wearable devices, and chip-on-film (COF) packaging. These technologies rely on the mechanical flexibility and durability of polyimides to create lightweight, adaptable designs.

4.5.5   Passivation and Encapsulation Layer Etching

Plasma Etching for Polymer Removal: Plasma etching is used to pattern polymer passivation layers, such as benzocyclobutene (BCB) or parylene. The precision of plasma etching ensures that only the necessary areas are etched, leaving other regions protected. This selective patterning is essential for exposing contact pads or interconnects without compromising the protective function of the passivation layer.

Oxide Removal for Rework: In some cases, encapsulation or passivation layers need to be removed for rework or reconfiguration. Wet etching can be used to dissolve oxides or other coatings, allowing engineers to make necessary adjustments to the device structure.

4.5.6   Etching in MEMS Packaging

MEMS combine mechanical structures with electronic circuits, requiring highly precise etching to form components like cantilevers, diaphragms, and interconnects. MEMS devices are increasingly integrated into modern packaging, especially in sensors and actuators.

DRIE for MEMS Structures: DRIE plays a pivotal role in the fabrication of MEMS devices, enabling the creation of deep cavities, channels, and other mechanical features with high aspect ratios. The precision offered by DRIE ensures that MEMS structures function reliably, meeting the stringent performance requirements of sensors and actuators.

Plasma Etching for Surface Cleaning: Plasma etching is used in MEMS packaging to remove surface residues or organic films that may interfere with the mechanical or electrical performance of the devices. This cleaning step ensures that MEMS components operate smoothly and efficiently.

4.6   Etching-Based Solutions for Critical Vulnerabilities in Microelectronics

Etching plays a vital role in addressing critical vulnerabilities that can compromise the performance, reliability, and security of microelectronic systems. As devices become increasingly complex, challenges such as signal interference, surface defects, adhesion issues, and environmental hazards require precise solutions. Through both wet and dry etching, engineers can mitigate these vulnerabilities, ensuring that systems function optimally under demanding conditions.

One significant vulnerability that etching addresses is signal interference, which can arise in high-frequency or dense interconnect designs. Improper routing of RDLs can lead to crosstalk and electromagnetic interference (EMI) between traces. Dry etching techniques, such as RIE, provide the precision needed to create finely defined metal traces with sharp profiles. This minimizes lateral spreading and improves the isolation between signals, which is essential for maintaining high signal integrity. In addition, dielectric layers can be precisely etched to further insulate signal paths, ensuring that high-frequency signals remain unaffected by interference. This level of control is critical in advanced packaging technologies, such as RF and millimeter-wave applications, where even minor variations can significantly impact performance.

Surface defects and contamination present another vulnerability that can affect bonding strength, electrical connections, and overall device reliability. Native oxide layers, residues from fabrication processes, or organic contaminants left from lithography steps can hinder adhesion or disrupt electrical continuity. Wet etching solutions, such as hydrofluoric acid (HF), are commonly used to remove native silicon dioxide layers, leaving behind a clean, bondable surface. Similarly, plasma etching is effective in removing organic residues, ensuring surfaces are free from contaminants that could compromise solder joints or interconnects. These processes play a crucial role in preventing electrical failures and mechanical defects, particularly in applications requiring robust interconnections over time.

The reliability of TSVs in 3D packaging presents another area where etching is indispensable. TSVs are vertical interconnections that pass through silicon wafers, enabling communication between stacked chips. However, mechanical stress and void formation within the vias can compromise their structural integrity. DRIE is essential for TSV fabrication, providing the precision needed to create deep, high-aspect-ratio vias with smooth sidewalls. The alternating passivation cycles employed in DRIE prevent void formation during subsequent metallization, reducing the risk of electrical failures. Smooth, vertical TSVs produced through DRIE also distribute stress evenly, mitigating the risks of cracking or delamination under thermal cycling conditions, which is especially important for high-reliability applications.

Adhesion between different layers of materials is another common vulnerability in microelectronic packaging, especially in flexible and multi-layer systems. Delamination or weak bonding can lead to mechanical failure, reducing the durability of the device. Plasma etching can enhance adhesion by modifying the surface energy of materials, creating a more receptive interface for bonding. For instance, plasma treatments can activate the surface of polymers or metal layers, improving their interaction with adhesives or coatings. Additionally, light etching of metallic layers can introduce micro-texturing, which increases surface roughness and enhances mechanical interlocking between bonded components. This approach ensures that multi-layer systems remain structurally stable under stress, making it particularly valuable for flexible electronics and wearable devices.

In addition to structural and electrical vulnerabilities, microelectronic devices must also be protected from environmental hazards such as moisture, dust, and chemical exposure. Etching techniques help define precise openings in protective layers without compromising environmental integrity. Plasma etching is frequently employed to pattern encapsulation materials like benzocyclobutene (BCB) or parylene, allowing selective exposure of contact pads or interconnects while maintaining the overall protective barrier. Wet etching is also used to ensure the integrity of hermetic seals by removing oxide layers that could interfere with proper sealing. This level of control is essential in ensuring that devices remain operational and reliable in harsh conditions, such as those encountered in automotive, aerospace, or medical applications.

In summary, etching provides targeted solutions to critical vulnerabilities in microelectronics, addressing challenges related to signal interference, defect mitigation, adhesion, structural integrity, and environmental protection. By leveraging the strengths of both wet and dry etching, engineers can develop systems that are not only highly functional but also durable and reliable. As packaging and integration technologies continue to advance, etching will remain a cornerstone technique, enabling manufacturers to overcome emerging challenges and deliver secure, high-performance microelectronic solutions.

4.7   Future Trends in Etching

The transition to smaller feature sizes and more complex architectures, driven by the need for increased transistor density, is also influencing the future of etching. As the semiconductor industry moves beyond 5-nanometer nodes, etching processes must provide atomic-level precision to define features with minimal dimensional variability. ALE is emerging as a critical technology in this area. ALE enables layer-by-layer material removal, allowing unparalleled precision and control over feature dimensions. This technique will be crucial for etching nanoscale interconnects, trenches, and vias, ensuring that microelectronic devices can meet the performance and reliability requirements of the next generation.

The introduction of new materials and device architectures is also prompting innovations in etching technologies. Materials such as GaN and SiC, which are used in high-power and high-frequency applications, pose unique challenges due to their hardness and chemical resistance. Future etching processes must adapt to efficiently and precisely remove these materials without compromising the underlying structures. Selective etching is becoming increasingly important, especially as devices incorporate complex multi-material stacks with metals, dielectrics, and polymers. Selective processes that target specific materials without affecting others will enable more efficient fabrication workflows and improve yield.

Sustainability and environmental responsibility are gaining prominence across the microelectronics industry, influencing the direction of etching technologies. Wet etching processes, while effective, often involve hazardous chemicals that require stringent disposal methods. To address environmental concerns, the industry is exploring green etching techniques that minimize chemical usage and waste. Innovations include developing safer, more environmentally friendly etchants and improving chemical recycling systems. In dry etching, reducing energy consumption and developing processes with lower carbon footprints are becoming priorities. Plasma systems with higher efficiency, as well as alternatives like gas-phase etching with minimal byproducts, are under development to align with sustainability goals.

Finally, process automation and AI are transforming the way etching is conducted in advanced manufacturing environments. The growing complexity of microelectronics demands tighter process control to maintain consistency across increasingly smaller features. AI-driven algorithms are being integrated into etching systems to optimize process parameters in real time, enhancing precision and reducing variability. In addition, automated monitoring systems are improving defect detection and ensuring that critical structures remain within specification throughout production. This level of automation is essential for meeting the high yield and reliability requirements of modern semiconductor manufacturing.

4.8   Industry Vendors and the Global Etching Ecosystem

Etching is a critical process in semiconductor manufacturing, playing a key role in both device fabrication and advanced packaging. The increasing complexity of microelectronics requires highly sophisticated equipment that can execute precise etching across a wide range of materials and structures. Numerous global vendors have developed state-of-the-art tools tailored for various etching needs, ensuring both efficiency and accuracy in semiconductor production. This section explores the leading vendors in the etching industry, their specialized products, and the economic ecosystem supporting these technologies at a global level.

4.8.1   Lam Research

Lam Research is a global leader in the development of semiconductor manufacturing equipment, focusing extensively on etching technologies. Lam offers a comprehensive portfolio of tools designed for various applications, including dielectric etch, conductor etch, and etch solutions for advanced packaging.

One of their flagship product families, Coronus®, specializes in bevel etching to remove residual materials along wafer edges. These residues, if not eliminated, can become sources of defects, potentially disrupting the manufacturing process and reducing device yield. The Coronus system combines precise plasma-based control with protective deposition technology to safeguard wafer edges. For example, in advanced memory applications like 3D NAND, prolonged etching can damage substrates, reducing yield. By integrating protective layers at strategic stages, Coronus® ensures wafer edge stability, preventing defects during bonding or 3D packaging processes.

Lam's Reliant® product line focuses on supporting the etching requirements of emerging markets such as MEMS and power semiconductors. These devices often involve larger or more specialized features than traditional nodes, necessitating highly adaptable equipment. Reliant® solutions extend the lifecycle of semiconductor fabs by enabling them to handle these diverse requirements efficiently, addressing the unique challenges of specialty technologies and advanced packaging.

4.8.2   Applied Materials

Applied Materials provides a diverse range of semiconductor manufacturing tools, with precision dry etching systems that are essential in integrated circuit fabrication and packaging. One of their notable products is the Mirra CMP 200mm system, which combines chemical-mechanical planarization (CMP) with post-CMP cleaning to ensure optimal etch performance.

The Mirra CMP platform addresses two critical needs: excess material removal and surface flattening for subsequent etching layers. With multi-zone polishing heads and high-speed platens, the system ensures uniform planarization across various materials, including silicon, polysilicon, and tungsten. It also incorporates specialized cleaning tools like the Desica® cleaner with Marangoni™ vapor drying to eliminate slurry residues and watermarks. This integrated approach ensures superior etching performance, particularly in copper damascene processes, WLP, and MEMS fabrication.

4.8.3   Tokyo Electron Limited (TEL)

Tokyo Electron (TEL) offers cutting-edge etching systems that address the challenges posed by shrinking technology nodes and complex architectures. Their Tactras™ platform is a 300mm plasma etch system designed to enhance productivity and support a variety of etching processes, including high-aspect-ratio hole etching, dielectric etch, and back-end-of-line (BEOL) mask removal.

The Tactras™ system provides flexible customization for specific applications, making it particularly valuable for advanced technology nodes. Its modular design enables seamless adaptation to different etch requirements, such as deep trench formation, which is crucial for memory devices, or high-precision mask etching in compound semiconductor technologies.

4.8.4   ASM International

ASM International offers semiconductor process tools that complement etching, such as PECVD and ALD. These systems play a crucial role in surface preparation and thin-film deposition, which are integral to many etching workflows. Precise deposition enhances the uniformity of layers that will undergo etching, ensuring better control and performance during manufacturing.

4.8.5   Hitachi High-Technologies

Hitachi High-Tech Corporation provides specialized etching equipment that supports the fabrication of next-generation devices, particularly those below the 20nm node. These devices require precise double-patterning techniques, 3D architectures, and high-precision etching to ensure structural integrity.

The Conductor Etch System 9000 Series developed by Hitachi is designed to meet these advanced requirements. It supports complex etching processes needed for forming protective layers, multi-material interfaces, and feature finishing, ensuring precise control over the geometry and surface properties of modern devices.

4.8.6   Plasma-Therm

Plasma-Therm specializes in plasma-based etching systems that address both dielectric and metal etching needs, with applications spanning MEMS, advanced packaging, and compound semiconductors. Their VERSALINE platform supports various etch and deposition processes, with advanced control systems like EndpointWorks® to enhance precision.

Plasma-Therm's ion beam technology is particularly valuable for processes requiring low-damage etching, such as deep silicon etching for MEMS and high-aspect-ratio etching for advanced packaging. Their systems also feature automated maintenance scheduling and remote monitoring capabilities, improving productivity and minimizing downtime in production environments.

4.8.7 Oxford Instruments

Oxford Instruments provides versatile tools for both research and production environments, with a focus on plasma etching. Their PlasmaPro 100 RIE modules offer both isotropic and anisotropic etching, making them suitable for a wide range of semiconductor applications. With load-lock and cassette-to-cassette options, these systems improve process repeatability, which is critical for achieving uniform etching in both research and high-volume manufacturing.

4.8.8 SPTS Technologies (an Orbotech Company)

SPTS Technologies specializes in wafer processing equipment, including etch, deposition, and thermal processing systems. Their Omega® etch systems are widely recognized for silicon DRIE, supporting applications such as wafer thinning, via formation, and MEMS fabrication.

The Rapier module within the Omega® platform utilizes the Bosch process for high-aspect-ratio silicon etching, enabling non-switched etching for tapered profiles and via reveals. These capabilities are essential for next-generation RF devices, MEMS sensors, and advanced packaging technologies, where precise control over etched features is required.

The Global Etching Ecosystem The semiconductor industry relies heavily on a robust ecosystem of vendors that provide etching equipment tailored to various manufacturing needs. These tools are crucial for producing devices across multiple markets, including logic chips, memory, MEMS, RF modules, and power semiconductors.

The economic impact of the etching ecosystem extends beyond equipment manufacturers to include material suppliers, service providers, and end-users. As device geometries shrink and packaging technologies evolve, demand for advanced etching solutions will continue to grow. This trend is reflected in the increasing investment in R&D by major vendors, as well as collaborations between equipment suppliers and semiconductor manufacturers to co-develop next-generation etching technologies.

The economic growth indicators for the etching market and the distribution of vendor contributions to global production show sustained growth in the sector. With the advent of emerging technologies such as 3D integration, heterogeneous packaging, and atomic layer etching, the etching ecosystem is poised for further expansion, driven by innovations in both tools and processes. The development of sustainable etching solutions will also become a priority, as the industry strives to balance technological advancement with environmental responsibility.

Overall, the global ecosystem of etching equipment vendors is a critical enabler of innovation in semiconductor manufacturing and packaging. Leading companies such as Lam

Research, Applied Materials, TEL, and others are driving the development of advanced etching technologies, ensuring that manufacturers can meet the evolving demands of smaller nodes, complex architectures, and higher performance. As the industry continues to push the limits of technology, collaboration among equipment suppliers, semiconductor fabs, and material providers will be essential in shaping the future of etching and maintaining the competitive edge of the microelectronics sector.

Conclusion:

In this chapter, we examined the essential role of etching in semiconductor fabrication and advanced packaging. Etching processes—both wet and dry—are fundamental in defining critical structures such as RDLs, TSVs, and MEMS components. Industry leaders like Lam Research, Applied Materials, and TEL continue to develop innovative solutions to meet the challenges of high-precision manufacturing. With trends moving toward atomic-level etching, sustainable practices, and automation, etching remains a pivotal technology in advancing microelectronics, ensuring it will be integral to future innovations in high-performance devices and systems.

References

1. Gong, Tianjiao, et al. "Investigation of vapor HF sacrificial etching characteristics through submicron release holes for wafer-level vacuum packaging based on silicon migration seal." *Journal of Microelectromechanical Systems* 32.4 (2023): 389–397.
2. Sun, Ke, et al. "Investigation into Photolithography Process of FPCB with 18 μm Line Pitch." *Micromachines* 14.5 (2023): 1020.
3. Zhuang, Dejin, and J. H. Edgar. "Wet etching of GaN, AlN, and SiC: a review." *Materials Science and Engineering: R: Reports* 48.1 (2005): 1–46.
4. Nojiri, Kazuo. *Dry etching technology for semiconductors*. Cham: Springer International Publishing, 2015.
5. Wilkinson, Chris DW, Ligang Deng, and Mahfuzur Rahman. "Issues in etching compound and Si-based devices." *Japanese Journal of Applied Physics* 41.6S (2002): 4261.
6. Kim, Dae Sik, et al. "Atomic layer etching applications in nano-semiconductor device fabrication." *Electronic Materials Letters* 19.5 (2023): 424–441.
7. Xu, Ya-dong, et al. "P-40: TFT-LCD a-Si Wet Etch Technology." In *SID Symposium Digest of Technical Papers*, vol. 54, no. 1, 2023.
8. Arana, Leonel R., et al. "Isotropic etching of silicon in fluorine gas for MEMS micromachining." *Journal of Micromechanics and Microengineering* 17.2 (2007): 384.
9. Meng, Lingkuan, and Jiang Yan. "Mechanism study of sidewall damage in deep silicon etch." *Applied Physics A* 117 (2014): 1771–1776.
10. Riesgo, Bibiana Vogel Peres, et al. "Effect of hydrofluoric acid concentration and etching time on the adhesive and mechanical behavior of glass-ceramics: A systematic review and meta-analysis." *International Journal of Adhesion and Adhesives* 121 (2023): 103303.

11. Kareem, Aseel A., et al. "Effect of phosphoric acid chemical etching on morphological, structural, electrical, and optical properties of porous GaAs Schottky diodes." *Journal of Materials Science: Materials in Electronics* 34.19 (2023): 1456.
12. Xie, Yongze, et al. "A long service life silicon-doped laser-etched polyimide anode materials for high-rate lithium-ion battery." *Journal of Energy Storage* 73 (2023): 108988.
13. Liu, Fangyuan, et al. "Unraveling Anisotropic and Pulsating Etching of ZnO Nanorods in Hydrochloric Acid via Correlative Electron Microscopy." *ACS Nano* 17.13 (2023): 12603–12615.
14. Wilkinson, C. D. W., and M. Rahman. "Dry etching and sputtering." *Philosophical Transactions of the Royal Society of London. Series A: Mathematical, Physical and Engineering Sciences* 362.1814 (2004): 125–138.
15. Kozlov, Andrei, et al. "Reactive ion etching of x-cut LiNbO3 in an ICP/TCP system for the fabrication of an optical ridge waveguide." *Applied Sciences* 13.4 (2023): 2097.
16. Kuzmenko, V. O., A. V. Miakonkikh, and K. V. Rudenko. "Investigation of Fluorocarbon Film Deposition from Ar/CF4/H2 Plasma for the Implementation of the Atomic Layer Etching Process." *High Energy Chemistry* 57.Suppl 1 (2023): S100–S104.
17. Li, Ningxin, et al. "Advances in High-Aspect-Ratio Deep Reactive Ion Etching of 4H-Silicon Carbide Wafers." *Journal of Microelectromechanical Systems* (2024).
18. Sharma, Pooja, et al. "Plasma etching of polycarbonate surfaces for improved adhesion of Cr coatings." *Applied Surface Science* 637 (2023): 157903.
19. Murakawa, Shigemi, and James P. McVittie. "Mechanism of surface charging effects on etching profile defects." *Japanese Journal of Applied Physics* 33.4S (1994): 2184.
20. Varga, Ksenija, et al. "Ultra High Density RDL Patterning of High–Resolution Dielectrics by Maskless Exposure Technology for High Performance Computing and Artificial Intelligence." In *2024 IEEE 74th Electronic Components and Technology Conference (ECTC)*. IEEE, 2024.

Physical Vapor Deposition in Advanced Semiconductor Packaging

5

5.1 Introduction to Physical Vapor Deposition (PVD)

5.1.1 Definition of PVD

PVD is a thin-film deposition technique that involves the transfer of material from a condensed phase—solid or liquid—into a vapor phase, followed by condensation onto a substrate to form a thin film. The process typically occurs in a vacuum environment to ensure purity and control over film growth. PVD offers precision in depositing metals, semiconductors, and dielectric layers, making it essential for semiconductor fabrication and packaging. Unlike CVD, which relies on chemical reactions, PVD is a purely physical process, driven by evaporation or sputtering of the source material. The films deposited through PVD are crucial for creating interconnects, protective coatings, and barriers in modern electronics.

5.1.2 Historical Background and Evolution of PVD

The origins of PVD date back to the 19th century, when early scientists experimented with metallic coatings through evaporation. The first significant developments came with thermal evaporation techniques in the early 1900s, which were used for decorative coatings on optical lenses and mirrors. However, it wasn't until the mid-20th century, with the emergence of semiconductor technology, that PVD began to play a key role in microelectronics.

The semiconductor industry, propelled by the invention of the transistor and ICs in the 1950s and 1960s, required thin, conductive films for interconnects and other device features. Early PVD processes focused on thermal evaporation, but as devices grew smaller and more complex, limitations in film uniformity and step coverage became apparent. In response, sputtering techniques emerged in the 1970s, providing better control over thin-film deposition. The introduction of magnetron sputtering further enhanced PVD's efficiency

© The Author(s), under exclusive license to Springer Nature Switzerland AG 2025
N. Asadizanjani et al., *Introduction to Microelectronics Advanced Packaging Assurance*,
Synthesis Lectures on Engineering, Science, and Technology,
https://doi.org/10.1007/978-3-031-86102-4_5

by increasing ionization within the plasma, allowing for faster deposition rates and better material utilization.

Over the decades, PVD has evolved alongside advancements in microelectronics. As technology nodes continued to shrink—moving from micrometer to nanometer scales—the demand for precise, high-purity films grew. Multi-layered thin films, barrier layers, and transparent conductive oxides were introduced to meet new challenges in semiconductor manufacturing, such as reducing signal delay and controlling diffusion. The rise of 3D integration, MEMS devices, and advanced packaging technologies, including WLP and TSVs, further expanded the role of PVD.

5.2   Types of PVD Processes

PVD involves several methods, each with unique characteristics suited for different applications in semiconductor manufacturing a few of them could be visualized through Fig. 5.1. The most common PVD techniques are sputtering, thermal evaporation, and electron beam (E-beam) evaporation [1]. These methods differ in how the source material is transformed into vapor and deposited onto the substrate. Below, each technique is explained with a focus on its working principle, variations, and applications in microelectronics.

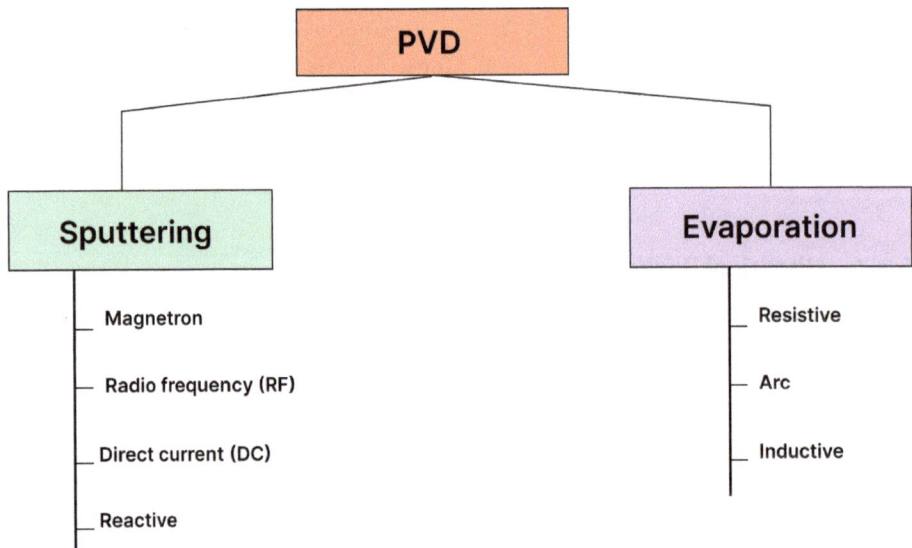

Fig. 5.1 Overview of physical vapor deposition (PVD) processes

5.2.1   Sputtering

Sputtering as illustrated in Fig. 5.2 is a widely used PVD process that relies on the physical ejection of atoms from a solid target material through ion bombardment [2]. In this method, an inert gas—commonly argon (Ar)—is introduced into the vacuum chamber, where it is ionized into plasma [3]. The positively charged ions from the plasma are accelerated toward the negatively biased target, striking its surface with enough energy to dislodge atoms. These ejected atoms then travel through the vacuum and condense on the substrate to form a thin film.

The underlying mechanism of sputtering involves momentum transfer. When ions bombard the target, energy is transferred from the ions to the atoms within the target material, causing them to escape from the surface. The vaporized atoms travel in straight lines toward the substrate, where they settle to form a uniform, thin layer.

5.2.1.1 DC (Direct Current) Sputtering

DC sputtering applies a continuous direct current (DC) voltage between two electrodes: the target material (cathode) and the substrate (anode). This voltage ionizes an inert gas, typically argon, creating a plasma. The positively charged ions within the plasma accelerate toward the negatively biased target, colliding with its surface and ejecting atoms that travel toward the substrate to form a thin film [4].

Working Mechanism of DC Sputtering The DC voltage creates a steady electric field that drives ion bombardment of the target material. As ions strike the target, momentum

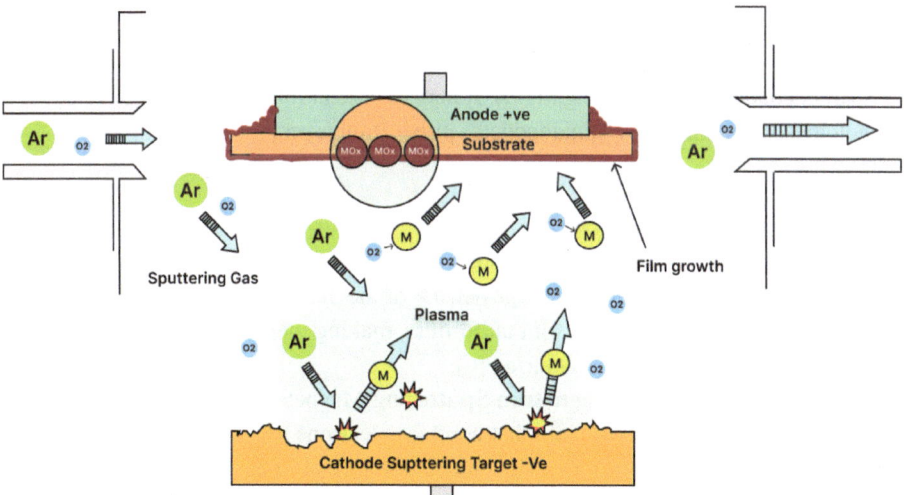

Fig. 5.2 An example of the processes involves in a sputter deposition system

is transferred, knocking atoms out of the surface. These ejected atoms travel through the vacuum and condense on the substrate. The deposition rate can be controlled by adjusting parameters such as gas pressure, voltage, and the distance between the electrodes.

Applications and Limitations DC sputtering is simple and cost-effective, making it ideal for depositing conducting materials, such as metals (e.g., copper, aluminum). However, it is not suitable for insulating materials, such as oxides or nitrides. When sputtering insulators, charge accumulates on the target surface, disrupting the plasma and halting the deposition process. DC sputtering is commonly used for applications where only metal films are required and where simplicity and speed are priorities.

5.2.1.2 RF (Radio Frequency) Sputtering

RF sputtering uses an alternating current (AC) at radio frequencies, typically 13.56 MHz, to power the plasma. This alternating current prevents the buildup of charge on the target material, making RF sputtering suitable for both conductive and insulating materials [5].

Mechanism of RF Sputtering In RF sputtering, the voltage oscillates between positive and negative values, switching polarity multiple times per second. This rapid alternation ensures that ions bombard the target during the negative half-cycle, while electrons neutralize any accumulated positive charge during the positive half-cycle. As a result, the plasma remains stable throughout the process, even when sputtering dielectric materials.

Advantages of RF Sputtering Material Versatility: RF sputtering can deposit both conductive (e.g., copper) and insulating materials (e.g., silicon dioxide).
Stable Plasma: The alternating polarity ensures that charge does not accumulate on the target, preventing plasma disruption.
Uniform Films: RF sputtering produces films with high uniformity, making it ideal for applications requiring precise thickness control.
RF sputtering is widely used to deposit materials such as dielectrics, semiconductors, and transparent conductive oxides (e.g., ITO). It plays a crucial role in applications like display technologies, solar cells, and semiconductor passivation layers.

5.2.1.3 Magnetron Sputtering

Magnetron sputtering is a variation of sputtering that incorporates magnetic fields around the target to trap electrons, increasing the density of the plasma. This enhancement leads to higher deposition rates and better film uniformity, making magnetron sputtering a popular choice for industrial-scale manufacturing [6].

Working Mechanism of Magnetron Sputtering Magnets are placed behind the target material to generate a magnetic field that confines electrons near the target surface. These trapped electrons increase the probability of ionization, creating a denser plasma close to the target. This higher plasma density improves the efficiency of ion bombardment, increasing the rate at which atoms are ejected from the target.

Because the plasma remains localized near the target, the substrate is exposed to fewer energetic ions, minimizing substrate damage and improving film quality. Magnetron sputtering systems are available in both DC and RF configurations, enabling the deposition of a wide range of materials.

Applications and Benefits High Deposition Rates: Magnetron sputtering allows faster material deposition, making it suitable for mass production.

Improved Film Quality: The localized plasma ensures uniform deposition with minimal defects.

Versatility: It can be used for both conductive and insulating materials, depending on whether a DC or RF power source is used.

Magnetron sputtering is extensively used in microelectronics for metallic thin films (e.g., copper, aluminum), dielectric layers (e.g., SiO_2), and barrier coatings (e.g., TiN). Its scalability makes it ideal for applications requiring large-area coating, such as in the semiconductor and display industries.

5.2.1.4 Reactive Sputtering

Reactive sputtering introduces a reactive gas, such as oxygen (O_2) or nitrogen (N_2), into the chamber along with the inert gas (argon). During the deposition process, the sputtered material reacts with the reactive gas, forming a compound film on the substrate. This method allows the deposition of materials that are difficult to sputter directly in their compound form, such as oxides, nitrides, and carbides [7].

Mechanism of Reactive Sputtering As ions bombard the target, atoms are ejected and react with the reactive gas in the chamber. For example, when titanium (Ti) is sputtered in the presence of nitrogen, the atoms combine to form titanium nitride (TiN) on the substrate. Similarly, introducing oxygen during aluminum sputtering forms aluminum oxide (Al_2O_3).

Reactive sputtering can also be controlled to deposit films with varying compositions by adjusting the ratio of inert to reactive gas. For example, the oxygen content in an oxide film can be tuned by regulating the oxygen flow during deposition.

Applications and Challenges Reactive sputtering is widely used to deposit functional thin films such as:

Titanium nitride (TiN): Used as a barrier layer in microelectronics.

Silicon dioxide (SiO_2) and silicon nitride (Si_3N_4): Essential for dielectric layers.

Aluminum oxide (Al_2O_3): Applied for protective coatings in optics and electronics.

One challenge in reactive sputtering is maintaining plasma stability, as introducing reactive gases can alter the plasma's behavior. Excessive reaction at the target surface can also lead to the formation of a "poisoned target," where the compound builds up on the target, reducing deposition rates. Advanced control systems and feedback mechanisms are often employed to manage these challenges and ensure consistent film quality.

Each type of sputtering—DC, RF, magnetron, and reactive sputtering—offers distinct advantages and addresses specific challenges in thin-film deposition. DC sputtering provides a simple and effective solution for metallic films, while RF sputtering expands the process to include insulating materials. Magnetron sputtering enhances deposition rates and film quality through the use of magnetic fields, making it ideal for large-scale applications. Reactive sputtering further broadens the scope of PVD by enabling the deposition of compound materials with tailored properties. Together, these sputtering techniques provide the versatility and precision needed to meet the complex demands of modern microelectronics.

5.3　Evaporation

5.3.1　E-Beam Evaporation

E-beam evaporation uses a focused electron beam to directly heat the target material, causing it to evaporate. This process operates in a high-vacuum environment, yielding high-purity films with precise thickness control, ideal for advanced microelectronics and optical applications.

Mechanism of E-Beam Evaporation An electron beam, generated by an electron gun, is focused onto the target material. The high-energy beam transfers sufficient energy to vaporize the material, which then condenses onto the substrate as a thin film. The process offers excellent control over deposition rates and target usage.

Advantages of E-Beam Evaporation
High Purity Films: The high-vacuum environment produces films with minimal contamination.
Precision Control: Allows precise control over deposition rates and film thickness.
Material Versatility: Suitable for a wide range of materials, including high-melting-point metals and dielectrics. E-beam evaporation is commonly used in semiconductor and optical industries for applications where high-purity, controlled-thickness films are essential.

5.3.2　Resistive Evaporation

Resistive evaporation uses an electric current to heat a resistive element, such as tungsten or molybdenum, which holds the target material. As the current flows, it generates heat, causing the target material to vaporize and deposit onto the substrate. This method is straightforward, cost-effective, and well-suited for low-melting-point metals and materials.

Mechanism of Resistive Evaporation In this process, the electric current flows through a resistive filament that holds the target material. The filament heats up, transferring energy to the material until it evaporates. The vapor condenses on the substrate, forming a thin film.

Advantages of Resistive Evaporation
Cost-Effective: Uses simple equipment, making it budget-friendly for basic coatings.
High Deposition Rate: Quickly deposits materials, particularly low-melting-point metals.

Good Film Uniformity: Provides uniform coatings across large areas with suitable setup. Resistive evaporation is widely used for coating materials like aluminum, gold, and silver in applications ranging from electronics to decorative finishes.

5.3.3   Arc Evaporation

Arc evaporation relies on a high-current, low-voltage arc to vaporize the target material. This method generates a high-density plasma at the target, producing highly ionized vapor for dense and durable coatings. Arc evaporation is often used for hard coatings and wear-resistant applications.

Mechanism of Arc Evaporation The arc is initiated between the cathode (target) and an anode, creating a localized high-temperature plasma that vaporizes the target material. The ionized vapor then condenses onto the substrate, forming a highly adherent and dense film.

Advantages of Arc Evaporation

High Adhesion: Ionized particles create strong adhesion on the substrate.

Dense Coatings: Ideal for wear-resistant applications due to its ability to produce hard, dense films.

Efficient Material Use: The process efficiently vaporizes the target, minimizing material waste. Arc evaporation is frequently used to deposit hard coatings, such as titanium nitride (TiN), in applications like cutting tools and wear-resistant surfaces.

5.3.4   Inductive Thermal Evaporation

Inductive thermal evaporation heats the target material using an inductively generated electromagnetic field. This field creates eddy currents within the conductive target or crucible, which heats the material to its evaporation point, making it suitable for high-melting-point materials.

Mechanism of Inductive Thermal Evaporation An inductive coil generates an electromagnetic field that induces eddy currents in the target material, heating it until it evaporates. This non-contact heating method minimizes contamination and provides precise control over the deposition rate.

Advantages of Inductive Thermal Evaporation

High-Temperature Capability: Effective for evaporating high-melting-point materials, such as refractory metals.

Minimal Contamination: Non-contact heating reduces contamination risks from the heating source.

Controlled Deposition: Provides accurate control over the deposition rate, resulting in uniform coatings. Inductive thermal evaporation is widely used for high-purity applications, such as depositing refractory metals and optical coatings.

5.4   Process Steps in PVD

PVD process involves several well-defined steps to ensure high-quality thin-film deposition with precise control over thickness, uniformity, and material purity. Each stage plays a crucial role in achieving the desired film properties essential for semiconductor manufacturing [8].

5.4.1   Substrate Preparation and Cleaning

The first step in PVD is thorough cleaning and preparation of the substrate to ensure proper adhesion of the deposited film. Any contamination, such as organic residues, dust, or oxides, can compromise the film's quality and functionality.

Wet Cleaning: Substrates are immersed in chemical baths (e.g., acetone or isopropanol) to remove organic contaminants [9].
Plasma Cleaning: A low-energy plasma treatment may be used to remove surface oxides and activate the substrate for better adhesion.
Drying: Substrates are dried carefully to avoid any moisture residues that could affect deposition. Clean substrates ensure that the thin film adheres well and that no defects or electrical issues arise from surface contamination.

5.4.2   Loading Into the PVD Chamber

Once cleaned, the substrates are loaded into the PVD chamber. Depending on the process, multiple substrates may be loaded into a carousel or holder for batch processing. Proper alignment is essential to ensure uniform film deposition across all substrates. The chamber is then sealed to maintain a controlled environment [10].

Automated wafer handling systems are commonly used in semiconductor fabs to avoid contamination from human contact. These systems transfer wafers between chambers and vacuum environments to maintain process integrity [11].

5.4.3   Vacuum Creation and Process Environment Setup

The PVD process occurs in a vacuum environment to prevent contamination and ensure free movement of atoms toward the substrate.

Vacuum Pumps: The chamber is evacuated to a pressure as low as 10^{-6} to 10^{-9} torr, depending on the material and process requirements.

Inert Gas Introduction: For sputtering processes, an inert gas (typically argon) is introduced to generate plasma.

Environmental Monitoring: Sensors continuously monitor chamber pressure, gas flow, and other parameters to ensure consistent process conditions. Maintaining a stable vacuum environment is crucial for precise film growth, as contamination from atmospheric gases can affect the film's electrical and mechanical properties.

5.4.4   Film Deposition: Monitoring Thickness and Uniformity

During deposition, the atoms or molecules ejected from the source material travel through the vacuum chamber and condense on the substrate to form a thin film. The deposition rate and thickness are controlled by adjusting the power supplied to the source and the chamber pressure.

Thickness Monitoring: Real-time monitoring tools such as quartz crystal microbalances (QCMs) [12] or optical interferometers [13] are used to measure the film's thickness with nanometer-level precision.

Uniformity Control: The substrate may rotate, or multiple sources may be used to ensure even coverage. In some cases, the chamber's pressure and temperature are adjusted dynamically to maintain uniformity. Precise control over thickness and uniformity ensures that the deposited film meets the required specifications for electrical performance and reliability.

5.4.5   Cooling and Unloading

After deposition, the substrate must cool down to avoid thermal stress or warping. Rapid cooling can induce cracks or film delamination, so cooling is typically gradual.

Once the substrate reaches room temperature, it is unloaded from the chamber, inspected for quality, and transferred to the next processing step. Automated systems handle substrate unloading to maintain cleanliness and avoid contamination.

5.5   Materials Deposited by PVD

PVD is highly versatile and capable of depositing a wide range of materials essential for semiconductor fabrication, from metals to dielectrics and compound materials. Below are some of the most common materials used in microelectronics manufacturing through PVD.

Metals: Copper, Aluminum, Titanium

Copper (Cu): Copper is widely used for interconnects in integrated circuits due to its high electrical conductivity. PVD copper deposition plays a crucial role in forming the metal lines and vias that connect different components on a chip.

Aluminum (Al): Before the adoption of copper, aluminum was the standard material for interconnects. It is still used in some applications, such as power devices and RF components, because of its corrosion resistance and ease of deposition.

Titanium (Ti): Titanium is often deposited as a barrier layer to prevent the diffusion of copper into the surrounding dielectric materials. It is also used as an adhesion layer to improve the bonding between the substrate and other metal films.

5.5.1 Dielectrics

Silicon Dioxide (SiO_2): SiO_2 is one of the most commonly used dielectric materials in semiconductor manufacturing. It serves as an insulating layer between metal interconnects and as a passivation layer to protect devices from environmental damage.

Aluminum Oxide (Al_2O_3): Al_2O_3 offers excellent electrical insulation and chemical stability. It is often used in high-performance devices to provide a protective coating or as a dielectric in capacitors. Dielectric materials are critical in maintaining electrical isolation and controlling capacitance in integrated circuits.

5.5.2 Transparent Conductive Oxides (TCOs): Indium Tin Oxide (ITO)

Indium Tin Oxide (ITO): ITO is a transparent conductive oxide that is widely used in display technologies, such as LCDs and OLEDs, and in touchscreens. Its combination of optical transparency and electrical conductivity makes it ideal for applications where electrical connections must be made without obstructing light. PVD processes enable precise deposition of ITO films with uniform thickness, ensuring high-quality displays and touch interfaces [14].

5.5.3 Alloys and Compounds: Titanium Nitride (TiN), Tantalum Nitride (TaN)

Titanium Nitride (TiN): TiN is commonly used as a diffusion barrier to prevent the migration of copper into dielectric layers. It is also applied as a hard coating in tools and MEMS devices due to its excellent wear resistance.

Tantalum Nitride (TaN): TaN is used in high-performance microelectronics for its thermal and chemical stability. It serves as a barrier layer in copper interconnects and as a resistive layer in thin-film resistors. The ability to deposit alloys and compounds with precision allows for the creation of films with tailored electrical, thermal, and mechanical properties, which are essential for advanced microelectronic devices.

These materials and process steps demonstrate the versatility of PVD and its ability to meet the demanding requirements of modern semiconductor manufacturing. Each step, from substrate preparation to film growth, plays a critical role in ensuring that the final product meets industry standards for performance and reliability.

5.6 Applications of PVD in Semiconductor Manufacturing

5.6.1 Interconnect Formation

Interconnects are the metal pathways that link individual transistors and components within an integrated circuit, enabling the flow of electrical signals across the chip. As circuits grow smaller and more complex, precise metal deposition becomes critical. PVD plays a vital role in forming these interconnects by depositing thin films of copper (Cu) and aluminum (Al)—the most commonly used interconnect materials [15].

Copper Deposition: Copper is favored for interconnects due to its low electrical resistance, which minimizes signal delay and power loss, especially in sub-10 nm technology nodes. PVD ensures that copper is deposited uniformly across the wafer, maintaining electrical integrity even as circuit densities increase [16].

Aluminum Deposition: Aluminum remains widely used for specific interconnects, particularly in power devices and RF components, due to its good conductivity and corrosion resistance. PVD ensures that aluminum films are smooth and adherent, forming reliable interconnects [17].

To maintain uniformity and coverage, magnetron sputtering is often employed. This ensures precise deposition over large surfaces and within complex circuit layouts, minimizing variability and enhancing the performance of the IC.

5.6.2 Barrier Layers for Diffusion Control (e.g., TiN, TaN)

In semiconductor manufacturing, controlling the diffusion of metals is critical to ensuring device stability. When interconnect metals like copper diffuse into adjacent dielectric layers, they can degrade the electrical performance of the circuit. Barrier layers, deposited through PVD, create an impermeable boundary that prevents such diffusion.

Titanium Nitride (TiN): TiN serves as an effective diffusion barrier due to its chemical inertness and stability. It ensures that copper remains confined within the interconnect lines, maintaining signal integrity over time. TiN films deposited via PVD are also used to improve adhesion between the metal interconnects and dielectric layers.

Tantalum Nitride (TaN): TaN is employed in high-performance chips, such as memory and logic devices, due to its superior thermal and chemical stability. It acts as a barrier to copper diffusion, especially in high-temperature processes. TaN's excellent resistance to corrosion makes it suitable for long-term reliability, even under extreme operating conditions.

The precision provided by PVD ensures that these barrier layers are thin, uniform, and defect-free, which is essential for ensuring the longevity and performance of modern semiconductor devices.

5.6.3   Contact Metallization for MEMS and Power Devices

PVD plays a pivotal role in the metallization of MEMS and power devices, where electrical contacts must be reliable and durable. These thin metal layers are critical for forming contact pads, electrodes, and interconnects, ensuring that devices function efficiently under demanding conditions.

Gold (Au) Contact Layers: Gold is often used in MEMS and sensors due to its excellent electrical conductivity and resistance to corrosion. PVD ensures precise deposition of gold films on sensitive surfaces, such as MEMS actuators and sensors, maintaining functionality in applications like accelerometers and gyroscopes[18].

Titanium (Ti) and Aluminum (Al) Electrodes: Titanium and aluminum are used as contact layers in power devices due to their conductivity and resistance to environmental degradation. These PVD-deposited layers improve the efficiency of devices such as power transistors and rectifiers, which must handle high voltages and currents [19].

PVD's ability to deposit thin films with high purity and adhesion ensures that these contact layers perform reliably over time, even in high-temperature or corrosive environments.

5.6.4   Thin-Film Resistors and Capacitors

Thin-film resistors and capacitors are essential components in analog, RF, and digital circuits, where precise control over electrical parameters is critical. PVD enables the deposition of both resistive and dielectric films with exceptional accuracy [20].

Resistive Films: Materials such as tantalum nitride (TaN) and nichrome (NiCr) are deposited via PVD to create thin-film resistors [21]. These resistors provide stable resistance values over a wide temperature range, ensuring consistent performance in circuits that require precise current control, such as voltage regulators and amplifiers.

Dielectric Films for Capacitors: PVD is used to deposit SiO_2 (silicon dioxide) and Al_2O_3 (aluminum oxide) as dielectric layers in thin-film capacitors. These materials ensure high capacitance with low leakage currents, making them ideal for RF applications and power conversion circuits. Capacitors formed using PVD-deposited films are critical in signal filtering and energy storage applications.

The ability of PVD to precisely control film thickness ensures that these components meet the stringent requirements of modern electronics, enabling high-frequency operation and stable long-term performance.

5.6.5 Decorative Coatings in Consumer Electronics

In addition to functional applications, PVD is widely used for decorative coatings in consumer electronics, where aesthetics and durability are paramount. PVD allows the deposition of thin films of precious metals and durable alloys, enhancing both the appearance and resilience of device surfaces.

Gold, Silver, and Titanium Coatings: PVD coatings provide a sleek, polished appearance to smartphones, wearables [22], and laptops. These films are not only aesthetically pleasing but also offer scratch resistance and corrosion protection, extending the product's lifespan.

Durability and Aesthetic Appeal: PVD coatings resist wear and tear, making them ideal for high-end consumer electronics, where products are subject to daily use. They are also hypoallergenic [23], making them suitable for wearable devices like smartwatches and fitness trackers.

PVD's ability to produce thin, durable coatings ensures that manufacturers can meet both functional and aesthetic demands, creating devices that are not only reliable but also visually appealing.

5.7 PVD in Advanced Packaging

As the demand for higher performance, miniaturization, and integration increases in the microelectronics industry, advanced packaging techniques have become essential. These packaging approaches—such as WLP, TSVs, and flexible packaging—enable the assembly of multiple components with high-density interconnects while maintaining a small form factor. PVD plays a critical role in these technologies by providing precise, high-quality thin films that are essential for forming electrical connections, redistribution layers, and flexible circuits. Below, we explore PVD's key contributions to advanced packaging solutions.

5.7.1 RDLs for Wafer-Level Packaging (WLP)

RDLs are crucial for WLP, as they facilitate the rerouting of I/O connections from the chip's original contact pads to new locations with a wider pitch. This redistribution simplifies the connection between the chip and the package or PC, enabling fan-out designs and multi-die integration [24].

PVD is employed to deposit the metal layers used in RDLs, most commonly copper (Cu) and aluminum (Al). The precision and uniformity of PVD ensure that these metal layers meet the stringent electrical and mechanical requirements of advanced packaging. Achieving uniform deposition over large wafer surfaces is essential to prevent electrical resistance variations that could affect signal integrity and device performance.

High Conductivity for RDLs: Copper is often the material of choice for RDLs due to its superior electrical conductivity, which minimizes power loss and signal delay. PVD ensures that copper films are deposited consistently, preventing voids or defects that could disrupt the signal path [25].

Multiple Metal Layers: PVD allows the deposition of stacked metal layers, which are critical for multi-layer RDLs. These multi-layer designs improve the routing flexibility of the package, enabling higher pin counts and more complex interconnections.

Adhesion and Barrier Layers: To enhance adhesion and prevent the diffusion of metals, titanium (Ti) or titanium nitride (TiN) barrier layers are often deposited through PVD before the copper or aluminum RDL layers are applied.

RDLs are integral to FOWLP, a technology that enables thinner, higher-performance packages by eliminating the need for an interposer. PVD's ability to deposit precise, uniform films ensures reliable electrical connections in these high-density, low-profile packaging solutions.

5.7.2   Through-Silicon Vias (TSVs) and Vertical Interconnects

PVD plays a crucial role in the metallization of TSVs and its process flow is presented in Fig. 5.3, where metal films such as copper are deposited along the walls of the via to form low-resistance electrical connections. The challenge in TSV metallization lies in achieving conformal coverage along the sidewalls of the deep, narrow vias, which typically have high aspect ratios.

Magnetron Sputtering for Uniform Deposition: Magnetron sputtering is often employed in TSV metallization, as it ensures that the deposited metal adheres uniformly along the sidewalls and base of the via. Advanced PVD techniques use rotating wafers or angled targets to further enhance film coverage.

Barrier and Seed Layers: Before depositing the primary metal, PVD is used to apply barrier layers (e.g., TiN) to prevent copper diffusion into the silicon substrate. A copper seed layer is also deposited through PVD, providing a conductive surface for the subsequent electroplating process that fills the via with copper.

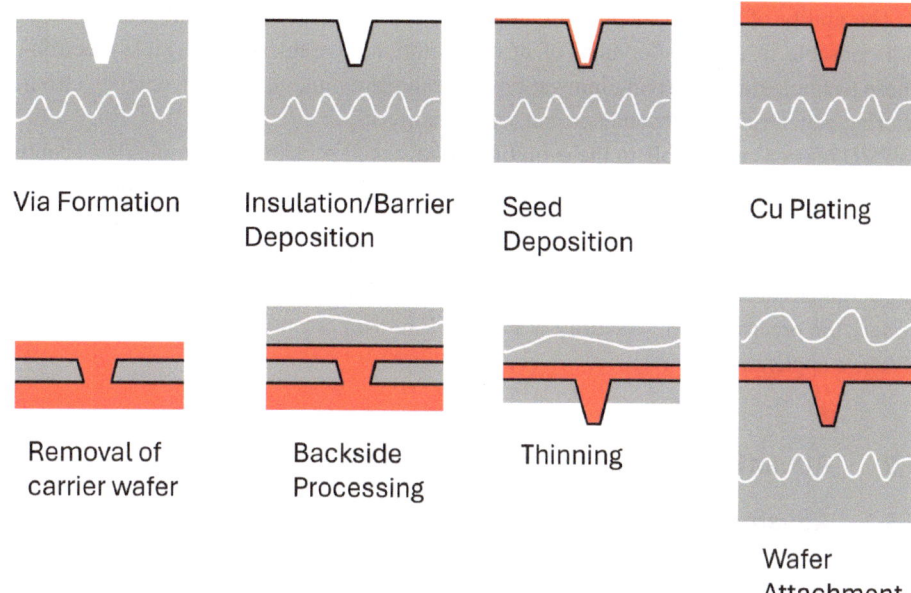

Fig. 5.3 Process Flow of Through-Silicon Via (TSV) Fabrication

5.7.3   Wearable Electronics and Flexible Packaging Applications

As electronics become more embedded in everyday life, the demand for wearable devices and flexible packaging has grown significantly. These devices require thin, lightweight, and flexible circuits that can withstand mechanical stress while maintaining electrical functionality. PVD is a key enabler for these applications, as it provides the precision needed to deposit conductive films and transparent electrodes on flexible substrates [26].

Conductive Films on Flexible Substrates: PVD is used to deposit metals such as gold, copper, and aluminum on flexible polymer substrates. These films serve as interconnects, electrodes, and sensor layers in wearable devices, such as fitness trackers, smartwatches, and biosensors [27].

Transparent Conductive Oxides (TCOs): Indium tin oxide (ITO), a transparent conductive material, is often deposited via PVD to create electrodes in flexible displays and touchscreens. ITO combines electrical conductivity with optical transparency, making it ideal for bendable displays and flexible sensors [28].

Mechanical Flexibility and Durability: PVD-deposited films must maintain their conductivity and adhesion under repeated bending and stretching. To achieve this, the deposition parameters are carefully controlled to minimize internal stress within the films, enhancing their mechanical flexibility and durability.

The ability of PVD to deposit high-quality thin films on flexible substrates makes it indispensable in the development of next-generation wearables and flexible electronics. These technologies are driving innovations in healthcare, fitness, and consumer electronics, offering new levels of convenience and functionality.

PVD plays a crucial role in the evolution of advanced packaging technologies, enabling higher levels of integration, performance, and reliability. By facilitating the precise deposition of metals and barrier layers in RDLs, TSVs, and flexible circuits, PVD ensures that electrical connections are robust, low-resistance, and scalable. As the industry moves toward more compact and complex packaging solutions, including 3D integration and wearable electronics, the precision and versatility of PVD will remain essential in meeting the demands of future applications.

5.8   Advantages and Challenges of PVD

5.8.1   Advantages of PVD

High Purity and Precise Thickness Control One of the most significant advantages of PVD is its ability to deposit films with high purity and minimal contamination. The process occurs in a vacuum environment, isolating the deposition process from atmospheric gases, dust, or moisture that could introduce impurities into the film. This high level of purity ensures that the electrical and mechanical properties of the deposited layers are consistent, which is essential for advanced semiconductor applications.

In addition to purity, PVD allows for precise control over film thickness. The deposition rate and final thickness can be finely tuned by adjusting parameters such as power, pressure, and target-substrate distance. Real-time monitoring tools, such as quartz crystal microbalances (QCMs) and optical interferometry, ensure that films meet exact specifications with nanometer-level precision. This precise control is especially important in applications like thin-film resistors, capacitors, and barrier layers, where even small deviations in thickness can affect performance.

Wide Range of Materials PVD supports the deposition of a broad spectrum of materials, including metals (e.g., copper, aluminum, titanium), dielectrics (e.g., silicon dioxide, aluminum oxide), oxides (e.g., indium tin oxide), and compounds (e.g., titanium nitride, tantalum nitride). This versatility makes PVD a suitable choice for fabricating various components within microelectronic devices, such as interconnects, electrodes, barrier layers, and transparent conductors.

The ability to deposit different materials allows engineers to tailor the properties of films for specific applications. For example, conductive layers can be used for interconnects, while dielectric coatings are ideal for insulating layers. Similarly, transparent conductive oxides (TCOs) are essential for touchscreens and displays, while hard coatings enhance the durability of consumer electronics.

Uniform Films with Excellent Adhesion Magnetron sputtering, produce highly uniform films with excellent thickness consistency across the substrate. Uniform deposition ensures that the electrical and mechanical properties of the film are consistent throughout the entire wafer, which is critical in applications like IC interconnects and wafer-level packaging.

PVD films also exhibit strong adhesion to the substrate, even under harsh operating conditions. The adhesion properties result from the kinetic energy of the atoms being deposited, which enhances the bonding between the film and the substrate. This makes PVD films highly resistant to delamination, even under thermal cycling or mechanical stress. As a result, PVD-deposited films maintain their integrity and functionality over the lifespan of the device.

5.8.2   Challenges of PVD

Vacuum Requirements and High Equipment Costs One of the primary challenges of PVD is the need for a high-vacuum environment to prevent contamination and ensure precise film deposition. The vacuum systems used in PVD, including pumps, gauges, and leak detection systems, are complex and require regular maintenance to maintain stability throughout the deposition process. The need for a cleanroom environment further increases the operational costs.

Additionally, PVD equipment is expensive to acquire and operate. Systems like magnetron sputtering setups and electron beam evaporators involve sophisticated controls, power supplies, and automated handling systems. These costs make PVD more suitable for high-value manufacturing environments, such as semiconductor fabs and advanced research facilities, but less practical for low-cost or large-volume production.

Poor Step Coverage for High-Aspect-Ratio Features A common limitation of PVD is its poor step coverage in high-aspect-ratio structures. Since the atoms or molecules being deposited travel in relatively straight paths through the vacuum, they may not reach deep into trenches, vias, or other intricate geometries. This can result in uneven coating, with thinner films along the sidewalls and at the bottom of these features.

This challenge becomes more pronounced in 3D packaging and TSVs, where the deposition of uniform metal layers along the entire structure is critical for electrical connectivity. In such cases, ALD or CVD may be preferred for better conformal coverage, though these techniques come with their own trade-offs in terms of speed and material selection.

Stress-Induced Defects in Thick Films PVD can introduce internal stress into deposited films, especially when thick layers are required. These stresses arise from factors such as thermal expansion mismatches between the substrate and the film, energy transfer during deposition, and residual strain within the film. If not properly managed, these stresses can lead to cracking, peeling, or delamination, compromising the performance and reliability of the device.

To mitigate stress-induced defects, engineers can adjust process parameters such as substrate temperature, deposition rate, and plasma power. Additionally, using multi-layer structures with alternating materials can help distribute stress and prevent film failure. Careful optimization is essential, especially in applications like MEMS devices and power electronics, where thick films are often required to provide durability and functionality.

5.9   PVD-Based Solutions for Solving Critical Vulnerabilities

In semiconductor manufacturing and advanced packaging, certain challenges, or vulnerabilities, can compromise device performance, reliability, and longevity. These vulnerabilities include issues such as electromigration, diffusion, poor adhesion, surface contamination, signal degradation, and mechanical stress. PVD offers targeted solutions to mitigate these vulnerabilities by providing high-purity films with excellent adhesion, precise thickness control, and versatile material deposition. Below are several ways in which PVD addresses these critical challenges.

5.9.1   Electromigration Mitigation for Interconnects

Electromigration occurs when high current densities cause atoms within metal interconnects to migrate, leading to void formation and, ultimately, circuit failure. As technology nodes shrink, the risk of electromigration increases due to smaller cross-sectional areas of interconnects.

Solution: PVD allows the deposition of high-purity copper (Cu) films, which are less prone to electromigration. Additionally, barrier layers made from titanium nitride (TiN) or tantalum nitride (TaN) can be deposited through PVD to prevent atom migration between metal lines and surrounding dielectrics. These thin, uniform PVD-deposited layers ensure that interconnects remain intact even under prolonged electrical stress, improving device reliability.

5.9.2   Diffusion Control with Barrier Layers

In advanced microelectronics, the diffusion of metal atoms, particularly copper, into dielectric materials can degrade device performance and lead to short circuits or signal loss. This diffusion challenge is critical in high-performance logic and memory devices.

Solution: PVD provides an effective solution by depositing impermeable barrier layers like TiN, TaN, or Al_2O_3. These films form a protective boundary around the metal interconnects, preventing the diffusion of atoms into adjacent materials. The precise thickness control provided by PVD ensures that the barrier layers are thin enough to maintain conductivity while still offering excellent diffusion resistance, preserving the performance and reliability of the circuit over time.

5.9.3 Enhanced Adhesion to Prevent Delamination

Poor adhesion between layers in a semiconductor device can lead to delamination [29], where films peel away under thermal or mechanical stress. This issue is especially problematic in MEMS devices and flexible electronics, where the films must withstand environmental variations and repeated mechanical deformation [30].

Solution: PVD addresses adhesion issues by enabling the deposition of adhesion-promoting layers such as titanium (Ti) and chromium (Cr). These thin layers are often deposited between the substrate and the primary material to enhance bonding. Additionally, the energetic nature of PVD ensures that the deposited atoms strongly adhere to the substrate, even in multi-layer structures, preventing delamination and ensuring long-term stability.

5.9.4 Surface Contamination Removal and Protection

Surface contamination, such as residual oxides or organic matter, can degrade electrical performance by interfering with conductivity or causing poor adhesion. Contamination is particularly problematic in contact metallization and solder bump formation during packaging.

Solution: PVD can be integrated with in-situ plasma cleaning to remove contaminants from the substrate surface before deposition, ensuring a clean surface for optimal film adhesion. Additionally, PVD can deposit protective coatings such as Al_2O_3 or SiO_2 to shield sensitive areas from future contamination, improving device reliability over time.

5.9.5 Signal Integrity Improvement for High-Frequency Applications

In high-frequency circuits, signal degradation caused by crosstalk, resistance, or poor conductivity can limit device performance. As circuit densities increase, maintaining signal integrity becomes more challenging.

Solution: PVD enables the deposition of high-purity copper (Cu) and transparent conductive oxides (e.g., ITO), which provide low-resistance pathways for electrical signals. Additionally, PVD-deposited metal-dielectric stacks [31] help to reduce electromagnetic interference (EMI) [32] and improve signal insulation. The precision of PVD ensures that the films are uniform and free from defects, which is essential for preserving signal quality at high frequencies.

5.9.6   Stress Management in Thick Films for MEMS and Power Devices

Internal stress in deposited films can lead to cracking, warping, or delamination, especially when thick layers are required for MEMS or power devices [33]. Stress-related defects can compromise the mechanical and electrical properties of the films, reducing device reliability.

Solution: PVD helps manage stress by enabling multi-layer deposition, where alternating materials are used to balance internal stresses. By carefully controlling the substrate temperature and deposition rate, PVD reduces residual stress within the films. The flexibility of PVD to deposit hard and soft layers allows for customized solutions that improve mechanical stability in devices subjected to extreme operating conditions.

5.9.7   Mechanical Flexibility for Wearable and Flexible Electronics

Wearable devices and flexible circuits must endure repeated bending, stretching, and twisting without losing electrical functionality. Traditional deposition techniques may result in films that crack or delaminate under mechanical stress.

Solution: PVD provides thin, conductive films with excellent adhesion and flexibility. Materials such as gold (Au), copper (Cu), and ITO are deposited through PVD onto flexible substrates like polymers, ensuring that the films maintain their conductivity even under mechanical deformation. PVD's ability to deposit uniform, high-quality films ensures that flexible circuits remain operational over long periods, making it ideal for wearable technology applications.

5.9.8   Prevention of Corrosion and Environmental Degradation

Electronic components are often exposed to moisture, dust, chemicals, and other environmental factors that can lead to corrosion or degradation of metal layers. This issue is especially critical in automotive, aerospace [34], and industrial electronics, where devices must operate reliably in harsh environments.

Solution: PVD allows the deposition of corrosion-resistant coatings such as titanium nitride (TiN) and aluminum oxide (Al_2O_3). These coatings protect metal layers from oxidation and corrosion, extending the lifespan of the device. PVD also enables the creation of passivation layers, which shield sensitive areas from environmental exposure and improve device durability.

5.9.9   Thermal Management for High-Performance Devices

As devices become more powerful, managing heat dissipation becomes critical. Poor thermal management can lead to thermal runaway, reducing performance or even causing device failure.

Solution: PVD helps deposit high-thermal-conductivity materials, such as copper and aluminum, which facilitate heat dissipation in high-power electronics. These films are used in thermal interfaces and heat spreaders to direct heat away from critical components [35]. Additionally, PVD-deposited reflective coatings can reduce thermal radiation, further enhancing thermal management.

5.10   Emerging Trends and Innovations in PVD

As semiconductor manufacturing advances, Physical Vapor Deposition (PVD) continues to evolve, with innovations that address the challenges of precision, material versatility, process efficiency, and sustainability. The latest trends in PVD focus on enhancing film quality, improving productivity through automation, and minimizing environmental impact. These developments ensure that PVD remains a critical technology for next-generation microelectronics and advanced packaging.

One significant innovation is High-Power Impulse Magnetron Sputtering (HiPIMS), which builds on traditional magnetron sputtering by delivering short, intense power pulses [36]. HiPIMS creates a dense plasma with a higher degree of ionization, resulting in improved film density and adhesion. This technique provides better control over film quality and ensures more uniform deposition, especially on complex surfaces with high-aspect-ratio features. HiPIMS reduces internal stress within the films, making it ideal for applications where mechanical stability is essential. Its ability to achieve superior film coverage and durability makes it particularly valuable in advanced packaging processes like TSV metallization and dielectric layer deposition.

Another critical development in PVD is the use of multi-source deposition systems, which enable the simultaneous or sequential deposition of different elements to create complex alloys and compounds. Traditional single-source PVD systems are limited to depositing pure elements, but multi-source systems allow for the precise formation of films with tailored electrical, mechanical, or optical properties. By fine-tuning the deposition rates of each source, manufacturers can create customized materials, such as transparent conductive oxides (TCOs) for displays and high-performance metallic alloys for semiconductor devices. This capability expands the range of materials that can be integrated into microelectronics, supporting innovations in logic, memory, and sensor technologies.

Process automation and AI-driven systems are becoming increasingly important in modern PVD applications [37]. Automated PVD systems feature robotic wafer handling and recipe management, ensuring consistency and reducing the risk of contamination. AI algorithms are also being integrated to optimize deposition parameters in real time, adjusting factors like power, gas flow, and chamber pressure to maintain film quality. Predictive maintenance enabled by AI minimizes downtime by identifying potential equipment failures before they occur. Automation and AI enhance productivity, improve film uniformity, and accelerate the development of new materials, making them indispensable in high-volume manufacturing environments like wafer-level packaging and MEMS production.

Sustainability is another focus area in PVD innovations, as the semiconductor industry seeks to reduce its environmental impact. PVD processes are inherently cleaner than chemical-based deposition techniques, but further advancements are improving energy efficiency and material recycling. Modern PVD systems incorporate energy-saving technologies that optimize vacuum efficiency and reduce power consumption during deposition. Additionally, efforts to recycle sputtering targets and recover precious metals, such as gold and platinum, are helping to reduce waste and operating costs. These eco-friendly practices align with the industry's push toward sustainable manufacturing, ensuring compliance with environmental regulations while maintaining production efficiency.

In conclusion, the latest trends and innovations in PVD such as HiPIMS, multi-source systems, AI-driven automation, and sustainable processes are shaping the future of microelectronics manufacturing. These advancements improve the quality, flexibility, and efficiency of thin-film deposition while addressing emerging challenges in advanced packaging and semiconductor fabrication. As technology continues to progress, PVD will remain a cornerstone of microelectronics, providing the precision and adaptability needed for the next generation of devices and systems.

5.11   PVD Ecosystem

The ecosystem surrounding PVD is highly complex, involving multiple companies, suppliers, and institutions working in tandem to produce the machinery, materials, and services required to deposit a single layer of thin film. The process of creating high-performance devices through PVD involves distinct stages: sourcing raw materials, designing and manufacturing specialized equipment, and finally, executing the deposition on wafers. Each step depends on contributions from various sectors, making the PVD supply chain a sophisticated network of equipment manufacturers, materials suppliers, component makers, and research institutions.

PVD technology plays an increasingly critical role in the semiconductor industry, and the demand for advanced equipment has grown as manufacturers seek to keep up with the scaling and performance requirements of next-generation microelectronics. As of 2020, the global PVD equipment market was valued at $19.35 billion, and it is expected to grow to

$25.93 billion by 2026, driven by the need for high-performance electronic devices. This growth is not only a result of the rising demand for chips and advanced packaging but also reflects the rising complexity and cost associated with developing cutting-edge PVD equipment and systems.

Geographical Market Dynamics The Asia-Pacific region holds the largest share of the global PVD equipment market, accounting for approximately 42% of the market in 2022. Countries such as China, Taiwan, South Korea, and Japan lead the way, benefiting from a well-established manufacturing infrastructure, substantial R&D investments, and the presence of major semiconductor companies. This dominance is also fueled by the cost advantages associated with production in these regions, including lower labor and material costs compared to other parts of the world.

5.11.1 Groups of PVD Equipment and Component Suppliers

The PVD equipment ecosystem is segmented into three main categories based on the roles companies play:

Large-Scale Equipment Manufacturers: These companies focus on developing and supplying industrial-scale PVD equipment for high-volume manufacturing (HVM). The most prominent company in this space is Applied Materials, which reported $26.5 billion in revenue last year. Applied Materials operates facilities across the globe, with its largest production facility located in Taiwan and other facilities in the United States, Germany, Israel, Korea, and Italy. Its PVD equipment is a cornerstone in both research laboratories and high-volume manufacturing plants worldwide, supporting the production of advanced semiconductor packaging and microelectronics.

Research and Development (R&D) Equipment Suppliers: This group focuses on smaller, specialized PVD equipment designed for research environments. R&D tools are critical for developing new materials and advanced deposition techniques that will eventually transition into industrial-scale production. These systems are generally used by universities, research institutions, and specialized labs working on new technologies.

Component Suppliers: Component suppliers provide key parts and materials essential to the operation of PVD systems. Companies such as Advanced Energy, ANCORP, and VacTechniche supply power supplies, target materials, vacuum components, and sealing technologies. These components are vital for ensuring the reliability and efficiency of PVD systems, playing an essential role in both R&D setups and high-volume production lines.

Conclusion:

This chapter explored the essential role of Physical Vapor Deposition (PVD) in semi-conductor manufacturing and advanced packaging. We examined key PVD processes, such as sputtering, thermal evaporation, and electron beam evaporation, along with their applications in interconnects, barrier layers, MEMS, and advanced packaging. PVD offers advantages like high purity, precise control, and broad material compatibility but also faces challenges such as vacuum requirements and stress defects. Emerging trends, including HiPIMS and AI-driven automation, continue to shape the field. As the industry advances, PVD remains pivotal in enabling high-performance, reliable, and sustainable electronics for the future.

References

1. Wang, Zhongping, and Zengming Zhang. "Electron beam evaporation deposition." *Advanced nano deposition methods* (2016): 33–58.
2. Hu, Boxuan, et al. "Advances in flexible thermoelectric materials and devices fabricated by magnetron sputtering." *Small Science* (2023): 2300061.
3. Carazzetti, Patrik, et al. "Study of Cross-Contamination in Multi-Chamber PVD Systems Used for High-Throughput Seed Layer Deposition." *2024 IEEE 10th Electronics System-Integration Technology Conference (ESTC)*. IEEE, 2024.
4. Kim, Sungtae, et al. "Boron-doped amorphous carbon deposited by DC sputtering for a hardmask: Microstructure and dry etching properties." *Applied Surface Science* 637 (2023): 157895.
5. Park, Chan-ho, et al. "Development of robust YSZ thin-film electrolyte by RF sputtering and anode support design for stable IT-SOFC." *Ceramics International* 49.20 (2023): 32953–32961.
6. Hu, Boxuan, et al. "Advances in flexible thermoelectric materials and devices fabricated by magnetron sputtering." *Small Science* (2023): 2300061.
7. Shapovalov, Viktor I. "Modeling of reactive sputtering—History and development." *Materials* 16.8 (2023): 3258.
8. Kaloyeros, Alain E., and Barry Arkles. "Silicon Carbide Thin Film Technologies: Recent Advances in Processing, Properties, and Applications: Part II. PVD and Alternative (Non-PVD and Non-CVD) Deposition Techniques." *ECS Journal of Solid State Science and Technology* 13.4 (2024): 043001.
9. Schwartz, Avital, et al. "Cleaning strategies of synthesized bioactive coatings by PEO on Ti-6Al-4V alloys of organic contaminations." *Materials* 16.13 (2023): 4624.
10. Carazzetti, Patrik, et al. "Study of Cross-Contamination in Multi-Chamber PVD Systems Used for High-Throughput Seed Layer Deposition." *2024 IEEE 10th Electronics System-Integration Technology Conference (ESTC)*. IEEE, 2024.
11. Hu, Hao, et al. "Prediction of wafer handling-induced point defects in 300 mm silicon wafer manufacturing from edge geometric data." *Solid State Phenomena* 345 (2023): 181–191.
12. Alanazi, Nadyah, Maram Almutairi, and Abdullah N. Alodhayb. "A review of quartz crystal microbalance for chemical and biological sensing applications." *Sensing and Imaging* 24.1 (2023): 10.

13. Kim, Sungtae, et al. "Design and assessment of phase-shifting algorithms in optical interferometer." *International Journal of Precision Engineering and Manufacturing-Green Technology* 10.2 (2023): 611–634.

14. Smirnov, Yury, Gaukhar Nigmetova, and Annie Ng. "Advances in Top Transparent Electrodes by Physical Vapor Deposition for Buffer Layer-Free Semitransparent Perovskite Solar Cells." *Solar RRL* (2024): 2400354.

15. Vorobyova, Mariya, et al. "PVD for Decorative Applications: A Review." *Materials* 16.14 (2023): 4919.

16. Soulié, Jean-Philippe, et al. "Cu1−xAlx films as alternatives to copper for advanced interconnect metallization." *Journal of Vacuum Science & Technology B* 42.4 (2024).

17. Kolehmainen, Jukka. "Aluminium Oxide Coatings for Electrical Insulation of Hard Carbon Thin Film Sensors." (2023).

18. Kottur, Himanandhan, et al. "Enhancing the MEMS Gyroscope Physical Assurance Using Quantum Sensing." *2024 IEEE RAPID Conference Proceedings.* IEEE, 1–2. 10.1109/RAPID60772.2024.10647048.

19. Persson, Eric. "Optimizing PCB layout for HV GaN power transistors." *IEEE Power Electronics Magazine* 10.2 (2023): 65–78.

20. Lu, Haidong, et al. "Domain dynamics and resistive switching in ferroelectric Al1−xScxN thin film capacitors." *Advanced Functional Materials* 34.28 (2024): 2315169.

21. Akkus, Meryem Sena. "Examination of the catalytic effect of Ni, NiCr, and NiV catalysts prepared as thin films by magnetron sputtering process in the hydrolysis of sodium borohydride." *International Journal of Hydrogen Energy* 48.60 (2023): 23055–23066.

22. Sun, Yuehang, Yun-Ze Li, and Man Yuan. "Requirements, challenges, and novel ideas for wearables on power supply and energy harvesting." *Nano Energy* 115 (2023): 108715.

23. Paul, Subhadeep, and Saikat Ghosh. "Nano-finishing of Natural Fibres." *Nanotechnology in Textile Finishing: Advancements and Applications.* Singapore: Springer Nature Singapore, 2024. 313–352.

24. Shen, Zesheng, et al. "Electromigration in three-dimensional integrated circuits." *Applied Physics Reviews* 10.2 (2023).

25. Gan, Chong Leong, and Chen-Yu Huang. "Advanced Memory and Device Packaging." *Interconnect Reliability in Advanced Memory Device Packaging.* Cham: Springer International Publishing, 2023. 1–19.

26. Wang, Yaxuan, et al. "Nano/micro flexible fiber and paper-based advanced functional packaging materials." *Food Chemistry* (2024): 140329.

27. Kim, Eun Ryung, et al. "Biosensors for healthcare: Current and future perspectives." *Trends in Biotechnology* 41.3 (2023): 374–395.

28. Dávila, Joel Arriaga, et al. "Enabling high-quality transparent conductive oxide on 3D printed ZrO2 architectures through atomic layer deposition." *Applied Surface Science* 636 (2023): 157796.

29. Šleiniūtė, Agnė, Gintaras Denafas, and Tamari Mumladze. "Analysis of the delamination process with nitric acid in multilayer composite food packaging." *Applied Sciences* 13.9 (2023): 5669.

30. Gu, Yue, and Yongjun Huo. "Advanced electronic packaging technology: From hard to soft." *Materials* 16.6 (2023): 2346.

31. Shi, Yuan-Kun, You-Ming Liu, and Hai-Feng Zhang. "The proposition of a dual-sensitivity laminated multi-metal dielectric stacks detecting structure based on the reflected Goos–Hänchen effect." *IEEE Sensors Journal* (2023).

32. Verma, Ritesh, et al. "A review on MXene and its' composites for electromagnetic interference (EMI) shielding applications." *Carbon* 208 (2023): 170–190.

33. Kottur, Himanandhan, et al. "Enhancing the MEMS Gyroscope Physical Assurance Using Quantum Sensing." *2024 IEEE RAPID Conference Proceedings*. IEEE, 1–2. 10.1109/RAPID60772.2024.10647048.
34. Soni, Rahul, et al. "A critical review of recent advances in the aerospace materials." *Materials Today: Proceedings* (2023).
35. Yussof, Amir Murtadha Mohamad, et al. "Revisiting the effectiveness of diamond heat spreaders on multi-finger gate GaN HEMT using chip-to-package-level thermal simulation." *Microelectronics Reliability* 161 (2024): 115496.
36. Oskirko, V. O., et al. "The influence of pulse duration and duty cycle on the energy flux to the substrate in high power impulse magnetron sputtering." *Vacuum* 216 (2023): 112459.
37. Aldoseri, Abdulaziz, Khalifa Al-Khalifa, and Abdelmagid Hamouda. "A roadmap for integrating automation with process optimization for AI-powered digital transformation." (2023).

The Crucial Role of CMP and Wafer Grinding

6

6.1 Introduction to CMP and Wafer Grinding

In semiconductor manufacturing, CMP and Wafer Grinding are essential processes that enable the fabrication of increasingly complex microelectronic devices. Both techniques play pivotal roles in ensuring that wafers and interconnect layers maintain precise flatness and surface integrity, crucial for the successful production of high-performance chips. As feature sizes shrink and devices require tighter interconnect pitches for higher performance, these processes become indispensable. They ensure that each layer within a microchip aligns correctly, preventing defects, enhancing yield, and enabling advanced packaging technologies such as 3D integration and heterogeneous packaging.

6.2 Definition of CMP and Wafer Grinding

CMP is a process that combines mechanical polishing with chemical reactions to achieve uniform planarization of surfaces [1]. It is employed to remove excess material from interlayer dielectrics [2], metals, or other thin films, ensuring the smooth, flat surfaces required for subsequent lithographic steps [3]. The technique uses a combination of abrasive slurry [4], polishing pads, and chemical agents to gently remove material in a controlled manner. This process is integral to the copper damascene [5] method and interlayer dielectric (ILD) polishing, enabling high-density interconnects in advanced nodes.

Wafer Grinding is a mechanical thinning process that uses abrasive grinding wheels to reduce the thickness of semiconductor wafers [6]. It is typically performed after wafer fabrication and before packaging and its role in packaging is depicted in Fig. 6.1, preparing wafers for backside processing or stacking in 3D ICs. Wafer grinding ensures precise thickness control, enhancing thermal dissipation, mechanical flexibility, and structural integrity.

© The Author(s), under exclusive license to Springer Nature Switzerland AG 2025   107
N. Asadizanjani et al., *Introduction to Microelectronics Advanced Packaging Assurance*,
Synthesis Lectures on Engineering, Science, and Technology,
https://doi.org/10.1007/978-3-031-86102-4_6

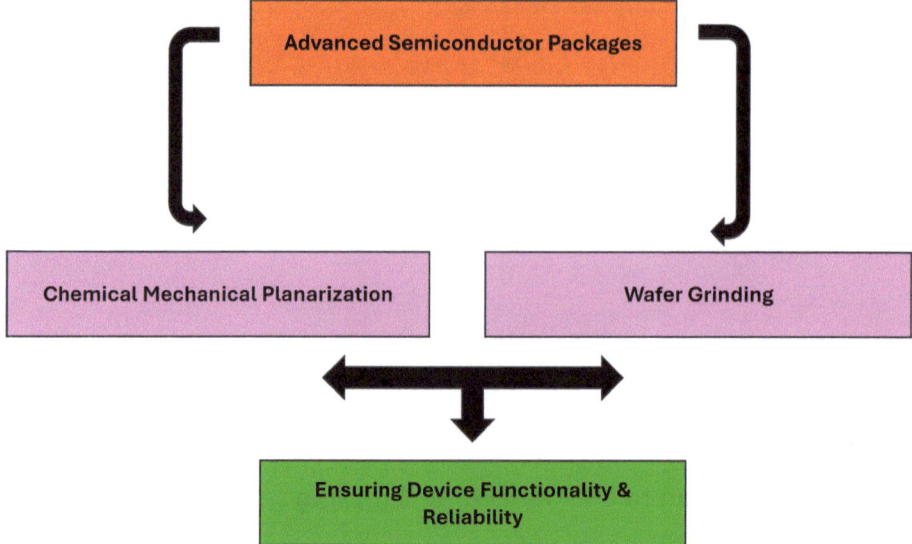

Fig. 6.1 Role of CMP and wafer grinding in advanced packaging fabrication

The process is essential for producing thin dies used in wearable electronics, FOWLP, and chiplet architectures [7].

6.3   Historical Evolution in Semiconductor Manufacturing

The origins of CMP and wafer grinding trace back to the early days of the semiconductor industry, where achieving flat surfaces for photolithography was already a concern. CMP was first introduced in the 1980s as a method to planarize ILDs for better lithographic patterning, particularly as feature sizes shrank. The rise of the copper damascene process in the 1990s further cemented CMP's importance, as it allowed the formation of copper interconnects with excellent electrical properties. Over the past decades, CMP has evolved into a highly specialized process, with custom slurries and polishing pads designed for specific materials like silicon dioxide, copper, tantalum [8], and aluminum oxide (Fig. 6.2).

Wafer grinding has a similarly long history, initially developed to thin wafers for power devices and sensors. With the advent of 3D ICs and advanced packaging, the demand for thinner, more flexible wafers has grown. As technologies such as TSVs emerged, wafer grinding became critical for ensuring precise thickness control and structural stability. In recent years, innovations in grinding tools [9], stress management techniques, and cooling systems have enhanced the precision and efficiency of wafer grinding, making it a key enabler of heterogeneous integration and chiplet packaging strategies.

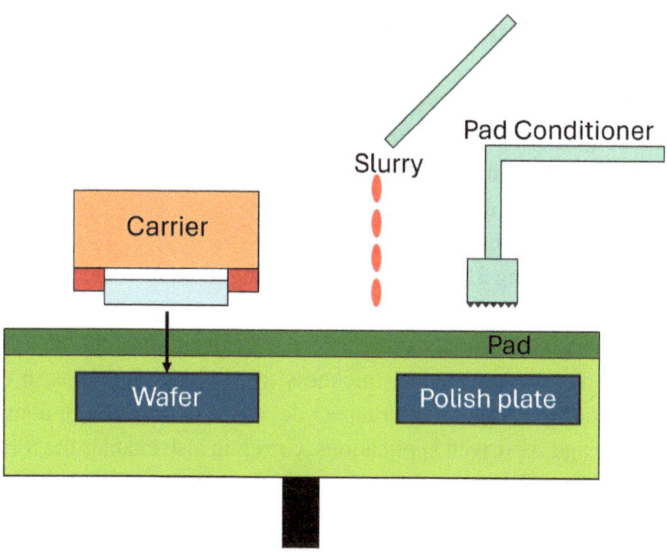

Fig. 6.2 Simplified diagram of the chemical mechanical planarization (CMP) Process

6.4   Importance of CMP and Wafer Grinding in Achieving Flat, Defect-Free Surfaces

Achieving flat, defect-free surfaces is essential for photolithography, interconnect formation, and device reliability. As chip geometries shrink to nanometer-scale nodes, even minor irregularities on the surface can result in patterning errors, signal interference, or yield losses. CMP plays a crucial role in eliminating surface topography variations, ensuring that metal layers and dielectrics are perfectly aligned for subsequent lithographic steps. Flat surfaces also reduce the risk of parasitic capacitance and signal distortion, improving electrical performance in high-frequency applications.

In wafer grinding, precise thinning is essential for producing ultra-thin wafers [10] that can withstand mechanical stress while maintaining thermal performance. These thin wafers are critical for 3D ICs, FOWLP, and wearable devices, where space and weight constraints are paramount. Grinding ensures consistent wafer thickness [11], minimizing the likelihood of warping or cracking during later stages of packaging and assembly [12]. It also improves the flatness of the wafer's backside, which is necessary for processes like TSV formation and wafer bonding.

6.5   Role of CMP and Wafer Grinding in Modern Fabrication

As the semiconductor industry moves toward advanced nodes and 3D integration, the importance of CMP and wafer grinding continues to grow. These processes enable the production of multi-layered structures with high-density interconnects, ensuring that each layer is perfectly planar and free from defects. CMP is essential for forming RDLs in wafer-level packaging (WLP), ensuring electrical connections are robust and reliable. It also ensures uniform TSV surfaces during 3D IC stacking, reducing the risk of electrical failures due to surface irregularities.

Wafer grinding plays an equally critical role in chip thinning for flexible electronics and 3D packaging. By reducing wafer thickness to just a few microns, it allows for the development of ultra-lightweight and compact devices that are in high demand for wearables, smartphones, and AI-driven applications. Grinding also enables the formation of thin dies that can be stacked efficiently, enhancing data transfer speeds and power efficiency in advanced systems such as AI accelerators [13], HBM, and network processors.

Together, CMP and wafer grinding form the backbone of modern semiconductor fabrication. These processes ensure the precise planarization and thinning required for scaling technology nodes, heterogeneous integration, and the development of next-generation devices. As chip architectures evolve, and the industry pushes toward smaller form factors and higher performance, CMP and wafer grinding will remain essential to overcoming technical challenges and ensuring product reliability.

6.6   Overview of CMP Process

6.6.1   Working Principle

Simultaneous Chemical and Mechanical Abrasion
The CMP process leverages chemical agents and mechanical abrasion [14] to remove unwanted material while achieving flat, uniform surfaces. A chemical slurry containing abrasive particles (such as silica or alumina) reacts with the target material to weaken the bonds at the surface. Simultaneously, a polishing pad applies mechanical pressure to the surface, grinding away the chemically softened material. The key to CMP's success lies in the balance between mechanical and chemical actions, ensuring that material removal is consistent across the wafer while minimizing surface defects. This combination enables high-precision planarization for materials like silicon dioxide, copper, tungsten as depicted in Fig. 6.3, and low-k dielectrics [15].

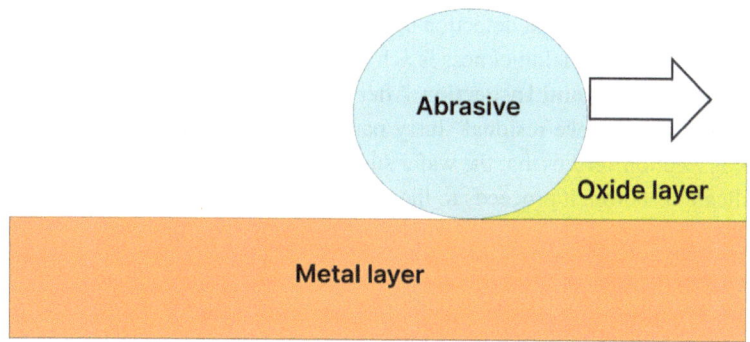

Fig. 6.3 Removal of surface oxide by abrasion in tungsten chemical mechanical polishing

6.6.2   CMP Consumables: Pads, Slurries, and Conditioners

CMP requires specialized consumables to perform effectively.

Polishing Pads: CMP pads are typically made from polyurethane [16] or composite materials. Their texture and hardness significantly impact material removal rates (MRR) and surface quality [17]. Pads must be carefully selected and maintained to ensure uniform abrasion without introducing scratches or defects.

Slurries: CMP slurries contain abrasive particles, oxidizing agents, and additives that react with the target material. The size and concentration of abrasive particles, as well as the slurry's pH, directly influence planarization efficiency and selectivity the ability to remove one material while leaving another intact.

Pad Conditioners: Over time, polishing pads degrade, reducing their ability to abrade material effectively. Pad conditioners [18], embedded with diamond particles, are used to refresh the surface of the pad, ensuring consistent performance throughout the CMP process.

6.6.3   Key Steps of CMP Process

Substrate Loading and Preparation The wafer is placed on a carrier head within the CMP tool, ensuring that it is securely held and aligned. Before starting the polishing process, the wafer undergoes pre-cleaning to remove any contaminants that might interfere with material removal or cause surface defects.

Application of Slurry and Pad Contact A controlled amount of CMP slurry is dispensed onto the rotating polishing pad, and the wafer is brought into contact with the pad. The carrier head applies pressure to ensure consistent contact between the wafer and the pad throughout the polishing process [19].

Mechanical Polishing with Chemical Reaction As the pad rotates, it applies mechanical abrasion to the wafer's surface, while the slurry chemically interacts with the target material to facilitate removal. End-point detection systems monitor the removal process in real time, ensuring the desired material thickness is achieved without over-polishing.

Post-CMP Cleaning and Inspection After polishing, the wafer undergoes a thorough cleaning process to remove residual slurry particles, contaminants, and polishing debris. Post-CMP inspection verifies that the wafer surface meets the required planarity and defect-free specifications before it proceeds to the next process step [4].

6.7   Overview of Wafer Grinding Process

Wafer grinding is a mechanical process used to reduce the thickness of semiconductor wafers, enabling the production of thin dies essential for 3D ICs, FOWLP, and wearable devices. It ensures that wafers meet precise thickness and flatness specifications, which are critical for modern chip designs that demand high performance, low power consumption, and compact form factors.

6.7.1 Working Principle

Mechanical Abrasion Using Grinding Wheels Wafer grinding relies on the use of diamond-coated grinding wheels to abrade the wafer's surface and remove material gradually. The wafer is secured on a vacuum chuck or with adhesive to prevent movement during grinding. The grinding process typically occurs in multiple stages to balance material removal speed and surface quality. This precision ensures that the wafer achieves its target thickness without warping or cracking, especially as wafers are thinned to just a few microns for stacked chips and flexible electronics.

6.7.2 Types of Grinding Processes

Back Grinding: Reducing Wafer Thickness for Packaging Back grinding is used to thin the backside of wafers after the front-side circuits have been fabricated and its emphasis in FOWLP is illustrated below in Fig. 6.4. This step is essential for advanced packaging techniques where thinner dies improve performance and allow for more efficient heat dissipation. Back grinding ensures that multiple dies can be stacked with minimal height, improving signal latency and power efficiency [1].

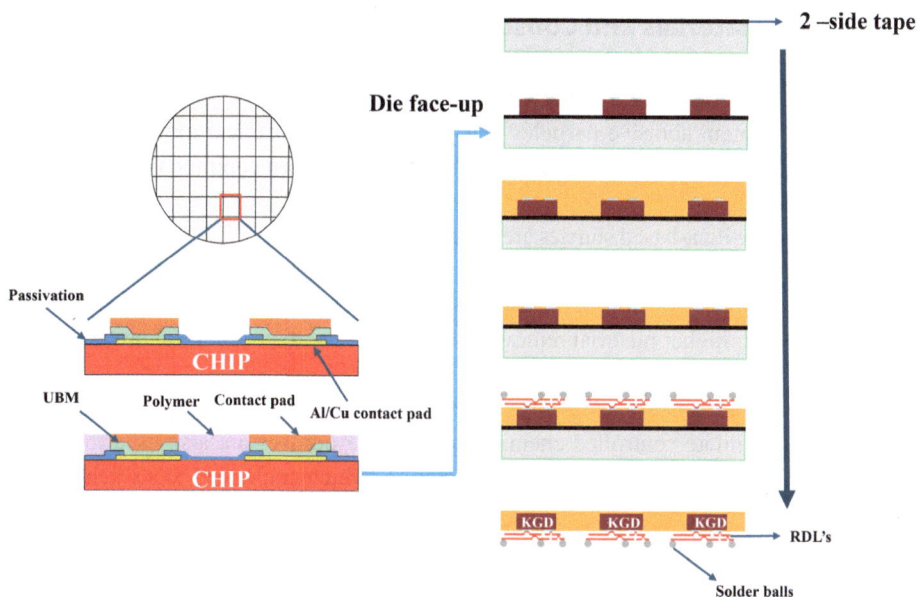

Fig. 6.4 Emphasis of backgrinding in Fan-Out wafer level (FOWLP) packaging

Edge Grinding: Preventing Wafer Cracking and Improving Strength Edge grinding smooths the wafer's periphery, eliminating any sharp or rough edges that could cause cracks during processing. Cracks that originate at the wafer's edge can propagate, reducing yield and reliability. Edge grinding also improves the mechanical strength of the wafer, ensuring it can withstand thermal cycling and handling during subsequent processes.

6.8   Fine Grinding Versus Rough Grinding: Surface Finish Control

Rough grinding is the first step, involving rapid material removal to reduce the wafer to a near-target thickness. This stage prioritizes speed but may leave the wafer surface slightly rough. Fine grinding follows rough grinding to improve the surface finish and bring the wafer to its final thickness specification. Fine grinding minimizes surface damage, ensuring that the wafer is smooth and defect-free for further processing, such as polishing, bonding, or TSV formation [20].

6.9   CMP Materials and Consumables

Slurry Chemicals: Silica and Alumina-Based Abrasives

CMP slurries contain abrasive particles that perform the mechanical removal of material while chemicals assist with oxidation and etching. The two most common types of abrasive materials are (Fig. 6.5).

Silica (SiO_2): Silica-based slurries are widely used for polishing ILDs and (SiO_2) layers. Silica particles provide precise material removal with minimal surface damage.

Alumina (Al_2O_3): Alumina slurries are used for harder materials, such as tungsten and copper. They offer higher material removal rates (MRR) but require careful process control to avoid scratches or surface defects.

In addition to abrasives, slurries contain oxidizing agents (like hydrogen peroxide) and pH buffers to facilitate controlled chemical reactions. Additives are also used to disperse abrasive particles evenly and prevent agglomeration, which could lead to scratches on the wafer surface.

Polishing Pads: Polyurethane and Composite Pads

Polishing pads are essential in CMP to provide the mechanical interface between the wafer and the slurry. These pads come in different materials and textures to suit various CMP applications:

Polyurethane Pads: These pads offer excellent durability and abrasion resistance, making them ideal for high-volume production. They are commonly used for polishing copper and tungsten laye

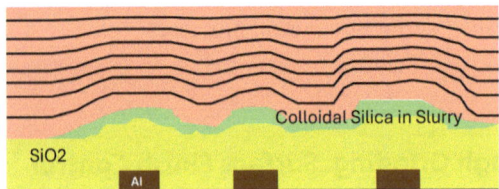

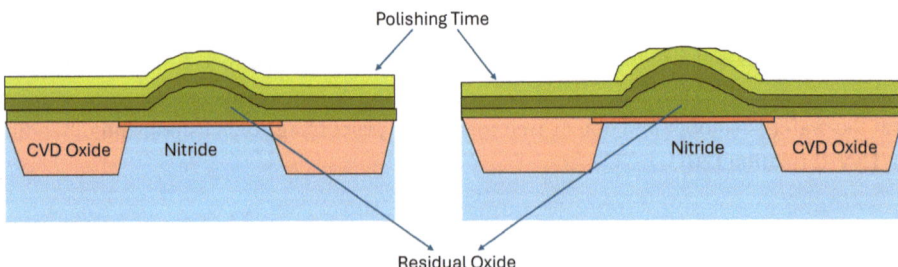

Fig. 6.5 Oxide planarization using chemical mechanical polishing (CMP)

Composite Pads: Composite pads combine polyurethane with other materials to create multi-layer structures that improve planarization performance. These pads are often used in polishing delicate low-k dielectrics and sensitive layers that require precise control over material removal.

Post-CMP Cleaning Agents After the CMP process, residual slurry particles and contaminants must be removed to prevent surface defects in subsequent steps. Post-CMP cleaning agents include specialized chemicals that remove abrasive particles without damaging the polished surface.

Surfactants and chelating agents help dissolve and lift slurry residues. pH buffers ensure the wafer surface remains chemically stable during cleaning. Post-CMP cleaning is a critical step to maintain the yield and reliability of advanced semiconductor devices.

6.10 Role of CMP and Wafer Grinding in Advanced Packaging Technologies

6.10.1 CMP in RDLs: Planarizing Metal Layers for Wafer-Level Packaging

In WLP, RDLs are used to extend the (I/O) connections of a chip to a broader pitch, making the chip compatible with external circuitry. The success of RDL formation relies heavily on the uniformity of metal layers. Copper and aluminum are typically used for RDLs, and any surface irregularities can disrupt electrical performance, causing signal degradation or short circuits.

CMP ensures a smooth and flat surface for each layer of metal in the RDL stack, improving yield and reliability. By removing surface irregularities, CMP allows for the seamless deposition of subsequent metal layers, minimizing defects and enhancing electrical performance. This process is particularly critical in FOWLP, where smaller form factors and high-density interconnects are essential for advanced applications.

6.10.2 CMP and TSV Processing: Ensuring Flat Surfaces for TSVs in 3D Packaging

TSV formation requires precise surface planarization to ensure smooth bonding between stacked layers and to prevent electrical failures.

CMP plays a vital role in the TSV process by polishing the top surfaces of the vias and surrounding dielectric materials to achieve perfect flatness and is depicted in Fig. 6.6. Any irregularities on the TSV surfaces can lead to misalignment, increased parasitic capacitance, and void formation, degrading the performance of 3D ICs. CMP ensures the smooth alignment of interconnects, enabling high-speed data transfer and improving the mechanical stability of stacked dies.

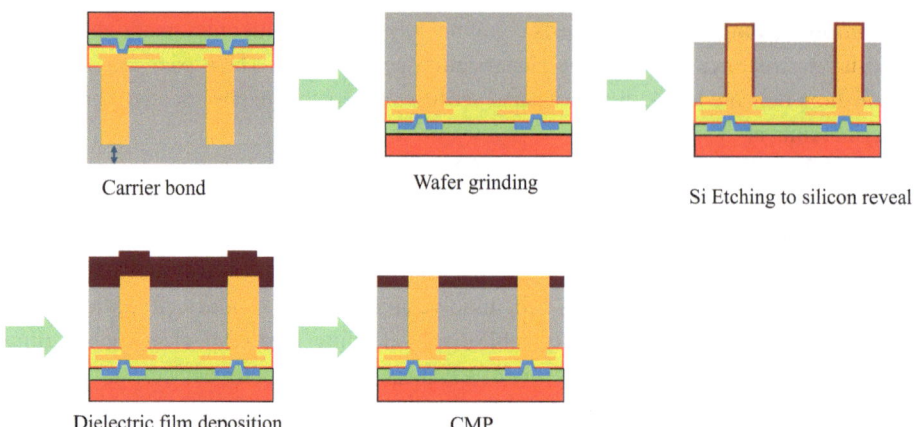

Carrier bond   Wafer grinding

Si Etching to silicon reveal

Dielectric film deposition   CMP

Fig. 6.6 Planarization of through-silicon vias (TSVs) Using chemical mechanical polishing (CMP)

6.10.3 Wafer Grinding for Flexible Electronics: Creating Ultra-Thin Wafers for Wearable Devices

Wafer grinding is crucial in producing ultra-thin wafers that are necessary for wearable electronics, medical devices, and flexible sensors. Modern applications demand lightweight, thin, and flexible chips that can operate reliably under physical stress.

Through back grinding, wafer thickness is reduced to just a few microns, enabling the creation of bendable substrates that maintain both mechanical flexibility and electrical performance. Wafer grinding ensures that these thin wafers can be stacked or integrated into compact packages without cracking or breaking, supporting the development of wearables, smart textiles, and next-generation displays.

6.10.4 Importance of Surface Finish Control in Chiplet and Heterogeneous Integration

Chiplet-based architectures and heterogeneous integration involve assembling multiple functional dies such as logic, memory, and sensors into a single package. Ensuring precise alignment and strong interconnects between these components requires an exceptional level of surface finish control.

CMP ensures that the surfaces of each chiplet and the interconnect layers are flat and defect-free, facilitating seamless bonding. This control over surface finish is critical for minimizing resistance, parasitic capacitance, and signal interference in high-performance devices. It also enhances the mechanical stability of heterogeneous packages, reducing the risk of failure during thermal cycling and device operation.

6.11   CMP-Based Solutions for Solving Vulnerabilities

6.11.1 Preventing Surface Defects and Contamination

During manufacturing, dishing, erosion, scratches, and particle contamination can compromise device yield and functionality. Surface defects, if left untreated, propagate through subsequent layers, leading to misalignment, reduced performance, and yield loss. CMP achieves uniform planarization, removing excess material while smoothing out topographical variations. This precision eliminates defects early in the process, ensuring that subsequent lithography and layer deposition occur without errors. By maintaining a smooth surface, CMP also minimizes contamination risks, reducing the chance of short circuits or open circuits caused by rogue particles.

6.11.2 Addressing Thermal Stress and Mechanical Fatigue

Devices operating under varying thermal conditions often experience stress due to differences in the thermal expansion coefficients of the materials used. This mismatch can lead to delamination, microcracks [21], or mechanical fatigue [22], compromising the structural integrity of advanced packages such as FOWLP. CMP ensures that RDLs and other interconnects are deposited with uniform thickness, reducing internal stress and preventing mechanical failure. This precision also helps devices maintain thermal stability over extended operational cycles, ensuring that packages can withstand thermal cycling without degradation.

6.11.3 Reducing Power Loss and Parasitic Effects in High-Density Interconnects

As device geometries shrink, surface irregularities become increasingly detrimental, causing leakage currents and increasing the resistance of metal lines. These parasitic effects degrade power efficiency, which is particularly problematic in low-power applications and RF circuits where signal integrity is critical. CMP addresses these challenges by delivering perfectly planar surfaces, minimizing the risk of leakage and ensuring that interconnects maintain low resistance. The enhanced smoothness also improves signal quality in high-density interconnects, supporting efficient power management in next-generation IoT devices, mobile processors [23], and AI accelerators.

6.12   Main Vendors in CMP and Wafer Grinding Technologies

Several companies lead the development of CMP and wafer grinding equipment, consumables, and technologies. These vendors provide solutions for high-volume production, offering tools that ensure precise planarization and wafer thinning for advanced packaging and 3D ICs.

6.12.1 CMP Vendors

The CMP ecosystem is dominated by companies that specialize in both polishing equipment and consumables, such as slurries and pads. Ebara Corporation is a key player, offering CMP machines for 300 and 200 mm wafers used in polishing materials like copper, tungsten, oxides, and other exotic materials. The company's polishing tools are widely used in high-end manufacturing, where precise planarization is critical for copper damascene processes and shallow trench isolation (STI).

In addition, Applied Materials offers a broad portfolio of CMP solutions, supporting a variety of applications from front-end semiconductor processing to niche technologies. Their equipment accommodates different wafer sizes and is compatible with leading-edge technology nodes as well as legacy processes. Companies like Fujimi and Fujifilm dominate the slurry space, supplying a wide range of formulations for polishing SiC, copper, aluminum, and other materials. DuPont and Entegris also play important roles, offering high-performance slurries and polishing pads for multiple substrate types.

6.12.2 Wafer Grinding Vendors

The wafer grinding industry benefits from overlapping expertise with CMP vendors, as many companies offer solutions for both processes. Ebara Corporation and Applied Materials leverage their experience in planarization to develop advanced wafer grinding machines that ensure thickness control and defect-free surfaces. However, the dominant player in this space is DISCO Corporation, a Japan-based company known for its innovative solutions in wafer thinning and dicing.

One of DISCO's key innovations is the "Dicing Before Grinding" (DBG) process, which allows the wafer to be singulated into individual dies before grinding is completed. This technique minimizes wafer breakage and backside chipping, enhancing the yield of thin wafers used in fan-out packaging and 3D ICs.

DISCO also offers a range of grinding wheels and consumables that support high-precision material removal. These grinding tools are designed to prevent static buildup and facilitate heat dissipation during grinding, ensuring superior surface quality. The ability to control wafer thickness precisely makes DISCO's equipment essential for producing thin, high-performance dies used in wearable devices and AI processors.

Conclusion:

In this chapter, we explored the essential role of CMP and wafer grinding in semiconductor manufacturing. CMP ensures smooth, defect-free surfaces through the precise interplay of chemical slurries, polishing pads, and machinery, enhancing device reliability and performance. Wafer grinding enables the production of ultra-thin wafers critical for advanced packaging, 3D integration, and flexible electronics. Key vendors such as Ebara, Applied Materials, and DISCO Corporation lead innovation in this field, ensuring that CMP and grinding technologies continue to meet the evolving demands of high-performance microelectronics. These processes remain foundational for achieving precision, efficiency, and scalability in modern semiconductor fabrication.

References

1. Herfurth, Norbert, et al. "Reliable backside IC preparation down to STI level using Chemical Mechanical Polishing (CMP) with highly selective slurry." *International Symposium for Testing and Failure Analysis*. Vol. 84741. ASM International, 2023.
2. Chouprik, Anastasia, et al. "Effect of Domain Structure and Dielectric Interlayer on Switching Speed of Ferroelectric $Hf_{0.5}Zr_{0.5}O_2$ Film." *Nanomaterials* 13.23 (2023): 3063.
3. Servin, Isabelle, et al. "Water-soluble bio-sourced resists for DUV lithography in a 200/300 mm pilot line environment." *Micro and Nano Engineering* 19 (2023): 100202.
4. Morinaga, Hitoshi. "Origin and Innovations of CMP Slurry." *ECS Journal of Solid State Science and Technology* 13.7 (2024): 074006.
5. Suzuki, Kenta, et al. "Electrical evaluation of copper damascene interconnects based on nanoimprint lithography compared with ArF immersion lithography for back-end-of-line process." *Japanese Journal of Applied Physics* 63.3 (2024): 03SP41.
6. Chiang, Tzu-Fan, et al. "3D grinding mark simulation and its applications for silicon wafer grinding." *The International Journal of Advanced Manufacturing Technology* 130.9 (2024): 4415–4430.
7. Zhou, Peipei, et al. "Refresh fpgas: Sustainable fpga chiplet architectures." *Proceedings of the 14th International Green and Sustainable Computing Conference*. 2023.
8. Lei, Rui, et al. "Interaction mechanism of Al_2O_3 abrasive in tantalum chemical mechanical polishing." *RSC Advances* 14.40 (2024): 29559–29568.
9. Song, Xin, et al. "Study on Preparation and Processing Properties of Mechano-Chemical Micro-Grinding Tools." *Applied Sciences* 13.11 (2023): 6599.
10. Chen, Yen-Shuo, et al. "Carbon/Nitrogen Dual-Doped in<100> P-Type Silicon Hard Mask for Wafer Thinning and Dishing Less for Hybrid Bonding." *2024 International Conference on Electronics Packaging (ICEP)*. IEEE, 2024.
11. Liu, Pan, et al. "The Impact of Wafer Warpage-Induced Unevenness on Alignment." *2024 Conference of Science and Technology for Integrated Circuits (CSTIC)*. IEEE, 2024.

12. Lei, Wei-Sheng, Ajay Kumar, and Rao Yalamanchili. "Die singulation technologies for advanced packaging: A critical review." *Journal of Vacuum Science & Technology B* 30.4 (2012).
13. Ortiz, Flor, et al. "Onboard processing in satellite communications using ai accelerators." *Aerospace* 10.2 (2023): 101.
14. Gherman, Raphael, et al. "Abrasive-free chemical-mechanical planarization (CMP) of gold for thin film nano-patterning." *Nanoscale* 16.36 (2024): 16861–16869.
15. Singh, Rajiv K., et al. "Fundamentals of slurry design for CMP of metal and dielectric materials." *MRS Bulletin* 27.10 (2002): 752–760.
16. Seo, Jangwon, et al. "Development of Novel Conditioning Method Using Thermal Shape Memory Characteristics of Polyurethane CMP Pad." *ECS Journal of Solid State Science and Technology* 13.3 (2024): 034003.
17. Kumar, Pankaj, et al. "Multiphysics simulation of the shape prediction and material removal rate in electrochemical machining process." *Advances in Materials and Processing Technologies* 10.2 (2024): 772–784.
18. Li, Xin, et al. "CMP Pad Conditioning Using the High-Pressure Micro-Jet Method." *Micromachines* 14.1 (2023): 200.
19. Park, Jeong-Heon, et al. "Study of over-polishing at the edge of a pattern in selective CMP." *Chemical Mechanical Planarization (CMP VI), S. Seal et al. Editors, The Electrochemical Society* (2003): 288–294.
20. Tao, Hongfei, et al. "Effects of grinding-induced surface topography on the material removal mechanism of silicon chemical mechanical polishing." *Applied Surface Science* 631 (2023): 157509.
21. Chen, Yuan, et al. "Research Status and Progress on Non-Destructive Testing Methods for Defect Inspection of Micro-Electronic Packaging." *Journal of Electronic Packaging* 146.3 (2024).
22. Xu, Xin, et al. "Fatigue behavior of 3D stacked packaging structures under extreme thermal cycling condition." *Memories-Materials, Devices, Circuits and Systems* 4 (2023): 100032.
23. Anitha, Subbarayan, Theagarajan Padma, and V. Vallimayil. "A generic resource augmentation architecture for efficient mobile communication." *Concurrency and Computation: Practice and Experience* 35.21 (2023): e7703.

Electrochemical Deposition in Advanced Packaging

This chapter presents a thorough investigation into electrochemical deposition (ECD) and its significance in the advanced packaging industry [1]. It explores the historical development and evolution of ECD, along with associated technologies, implementation methodologies, and the industrial framework. The chapter also addresses equipment suppliers, technical challenges, and potential avenues for further research and trend analysis within the field. ECD emerges as a pivotal technology, offering precise control for applying thin layers of conductive material, thus forming a cornerstone for technological advancements in advanced packaging.

7.1   Introduction to Electrochemical Deposition (ECD)

Electrochemical Deposition (ECD) has become an essential process in the manufacturing of modern semiconductor devices, enabling the precise deposition of metal layers that form critical interconnects, RDLs, and bumps used in advanced packaging. As the complexity of microelectronic circuits continues to increase with the demand for faster, smaller, and more efficient devices, ECD offers manufacturers a scalable solution to produce high-density interconnects with excellent electrical performance and reliability. This process plays a key role in facilitating 3D integration, FOWLP, flip-chip bonding, and heterogeneous integration, making it indispensable in semiconductor fabrication.

7.2   Definition of Electrochemical Deposition (ECD)

At its core, Electrochemical Deposition involves the controlled deposition of metal films onto conductive surfaces through electrochemical reactions. The target wafer or substrate serves as the cathode (negative electrode), while an anode (positive electrode) completes the electrical circuit. When an electric current is passed through an electrolyte solution containing dissolved metal ions, these ions migrate toward the cathode and form a thin metal film on the wafer's surface. Two main types of ECD are commonly used in semiconductor manufacturing:

Electroplating: A current-driven process where metal ions are deposited on the surface under an applied electric field.

Electroless Deposition: A chemical reduction process where metal layers are deposited without the need for an external current.

Both techniques have specific advantages, making ECD a highly flexible method for creating uniform, conductive layers in a wide variety of applications, including copper interconnects, TSVs, RDLs, and MEMS components.

7.3   Historical Development and Evolution of ECD in Semiconductor Manufacturing

While the fundamental principles of ECD trace back to the early 19th century, its application in semiconductor fabrication gained significant traction in the 1980s and 1990s, prompted by the industry's shift from aluminum to copper interconnects. Copper, with superior electrical conductivity and enhanced resistance to electromigration, became the preferred material. However, conventional deposition methods like PVD faced challenges in filling increasingly narrow trenches and deep vias. Electroplating, leveraging ECD principles, emerged as the ideal solution to fill high-aspect-ratio features with dense, void-free copper films, ensuring reliable interconnections.

The evolution of ECD in semiconductor technology has roots in earlier innovations, including IBM's magnetic memory technology in the 1960s. Lithography, combined with ECD, was initially employed to create memory structures by filling shaped patterns with conductive metals. Advancements continued through the 1970s with LIGA techniques as illustrated in Fig. 7.1 incorporated e-beam, x-ray lithography, and electroplating—enabling the formation of sub-micron features by the 1980s.

The 1990s and early 2000s marked a pivotal period where ECD research expanded to non-metallic substrates like ceramics, glass, and plastics. This era saw breakthroughs in electroplating multi-layer polyaniline on glass electrodes and creating metal nanoparticle films on glass and silicon. As ECD capabilities broadened, copper and zinc oxide layers were developed on silicon, while metal-organic frameworks emerged in 2010, enabling custom nanoscale structures with catalytic properties for specialized reactions.

Today, ECD is integral to both front-end processing and back-end packaging technologies, supporting the manufacturing of high-performance AI processors, automotive sensors, and 5G communication devices. ECD has expanded beyond copper to include metals like nickel,

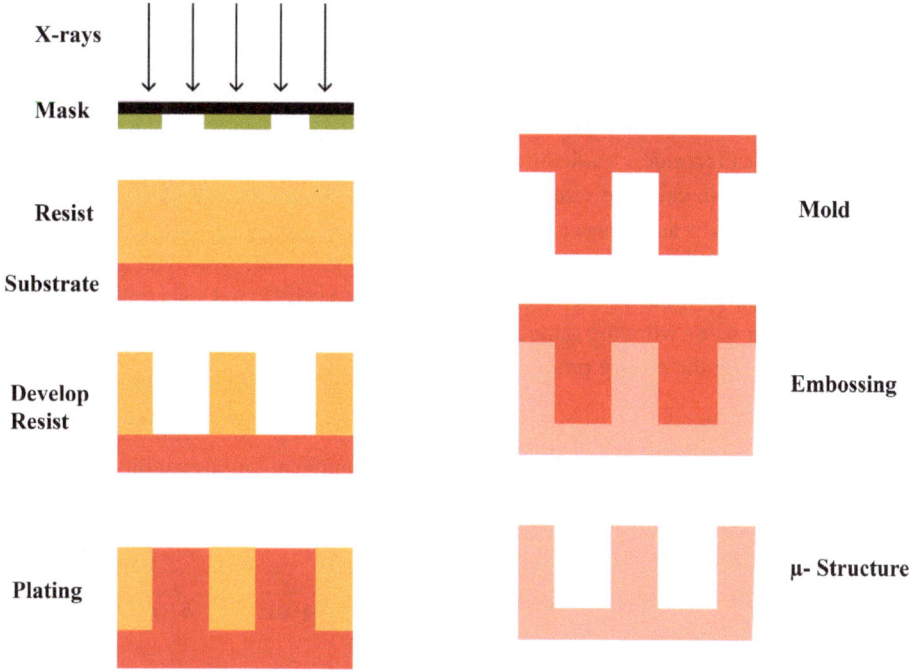

Fig. 7.1 LIGA process in electrochemical deposition (ECD)

gold, tin, and silver, serving applications in interconnects, MEMS devices, and protective coatings. The process has also become a cornerstone in advanced nodes, allowing precise deposition in complex architectures such as 3D ICs, TSVs, and FOWLP.

The drive for innovation continues as ECD research explores new materials and nanostructures to optimize performance and efficiency in diverse applications. For instance, in solar cell fabrication, ECD helps reduce costs and enhance manufacturing speed by integrating packaging and semiconductor production into a unified process. ECD remains a critical focus, motivated by the ongoing race to develop adaptable, efficient energy sources and cutting-edge semiconductor technologies.

7.4   Types of Electrochemical Deposition Processes

Electrochemical deposition encompasses two primary techniques: electroplating and electroless deposition. While both involve the deposition of metals onto a surface, they differ significantly in terms of the mechanisms and control parameters involved. Each method offers unique capabilities for depositing thin, uniform films, essential for high-precision microelectronics manufacturing.

7.4.1 Electroplating

Electroplating is a technique in which metal ions are deposited onto a conductive surface through the application of an external electric current [2]. The target surface acts as the cathode (negative electrode), while the anode (positive electrode) maintains the flow of ions in the electrolyte bath could be visualized in Fig. 7.2. Electroplating allows precise control over deposition rates by adjusting current density, electrolyte concentration, and plating time.

The basic mechanism involves the migration of metal ions through the electrolyte. At the cathode, these ions gain electrons and are reduced to form a solid metal layer on the substrate. The anode, often made from the same metal, dissolves into the electrolyte to replenish the metal ions, ensuring the plating process can continue uniformly over time.

Direct Current (DC) Plating vs. Pulse Plating DC electroplating applies a continuous current throughout the process. While effective for many applications, it can sometimes result

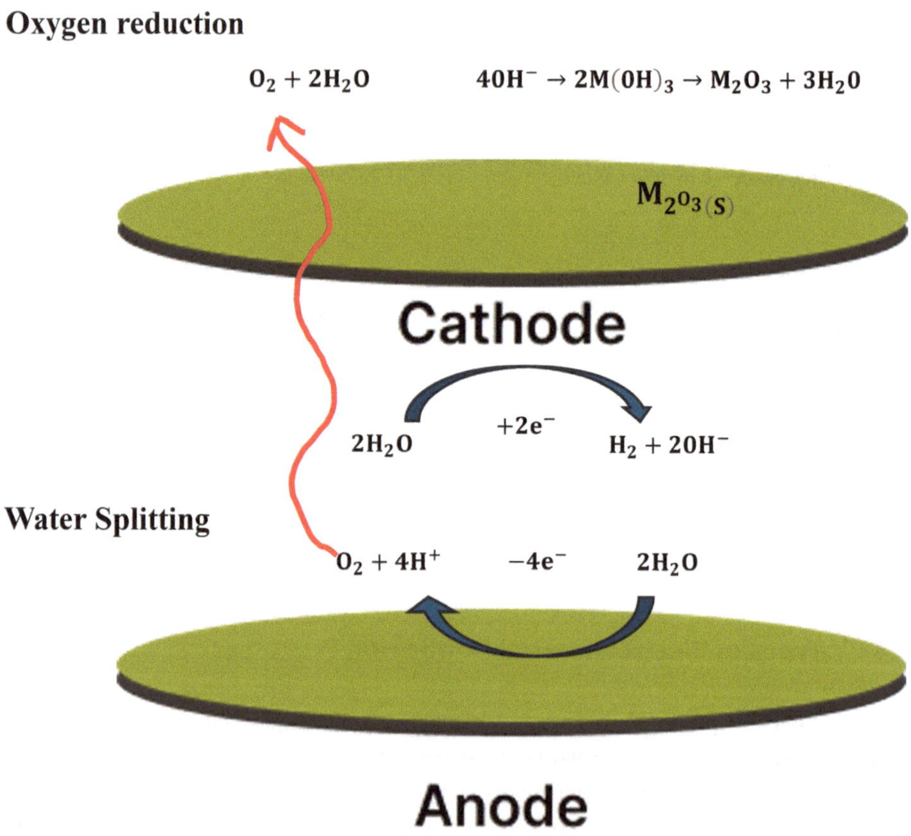

Oxygen reduction

$$O_2 + 2H_2O \qquad 4OH^- \rightarrow 2M(OH)_3 \rightarrow M_2O_3 + 3H_2O$$

$$M_2O_{3(S)}$$

Cathode

$$2H_2O \qquad +2e^- \qquad H_2 + 2OH^-$$

Water Splitting

$$O_2 + 4H^+ \qquad -4e^- \qquad 2H_2O$$

Anode

Fig. 7.2 Electroplating driven by redox reactions

in non-uniform film deposition, especially on complex surfaces or in deep vias and trenches. Uneven current distribution can lead to voids, rough surfaces, or inconsistent thickness.

Pulse plating improves upon traditional DC plating by applying current in intermittent cycles, alternating between plating and rest periods. This technique promotes more uniform ion distribution across the surface, reducing the risk of defects. During the off periods, metal ions can redistribute evenly, which helps avoid localized stress and grain growth anomalies. Pulse plating also enhances the microstructure of the deposited film, improving its mechanical properties and surface smoothness.

Mechanism of Electroplating Electroplating relies on the reduction of metal ions at the cathode to form a solid layer. The metal deposition is governed by Faraday's laws, which correlate the amount of material deposited with the applied current and time. An electric field facilitates the transport of metal ions from the electrolyte to the cathode, and the anode continuously dissolves to replenish the ion concentration in the bath. The electrolyte solution must remain stable and well-regulated to ensure uniform deposition and prevent chemical imbalances that can degrade film quality.

7.4.2 Electroless Deposition

Electroless deposition differs from electroplating in that it does not require an external power source [3]. Instead, it relies entirely on chemical reactions between a reducing agent and metal ions within the plating solution. The process is catalyst-driven—a thin catalytic layer, such as palladium, initiates the deposition, promoting the reduction of metal ions on the substrate's surface [4].

Electroless deposition offers unique advantages, particularly for surfaces with complex geometries or those that lack electrical conductivity. Since the process is purely chemical, metal deposition occurs uniformly across all exposed surfaces, even in narrow features where electric fields may not penetrate effectively in electroplating.

Mechanism of Electroless Deposition In electroless deposition, the metal ions are reduced by a chemical reducing agent (such as hypophosphite or formaldehyde), depositing a solid metal layer onto the substrate [5]. The presence of a catalyst ensures that the reduction occurs selectively on the desired areas, promoting consistent deposition without the need for external current. Temperature, pH, and bath composition are carefully controlled to ensure stable deposition rates and prevent unwanted precipitation of metals in the electrolyte [6].

The absence of an external current makes electroless deposition particularly suited for processes where even coverage is required over non-conductive surfaces or complex structures. However, maintaining the chemical stability of the bath is critical to avoid depletion of reactants or undesired side reactions. Proper control of bath conditions also ensures precise control over film thickness and grain structure.

These two electrochemical deposition methods—electroplating and electroless deposition—form the foundation of modern metal deposition technologies used in semiconductor manufacturing. While electroplating offers precise control through electrical parameters, electroless deposition provides a chemical-based alternative for achieving uniform coatings in challenging geometries. Each technique presents unique advantages, contributing to the flexibility and scalability required in advanced semiconductor processes.

7.5   Process Steps in Electrochemical Deposition (ECD)

The ECD process requires precise control over several steps to ensure high-quality metal films with excellent uniformity, adhesion, and performance. Each step plays a vital role in preparing the surface, establishing conductive pathways, controlling deposition conditions, and treating the deposited film to meet stringent semiconductor manufacturing standards. Below is a detailed discussion of the key stages involved in the ECD process.

The first step is surface preparation and cleaning, where the wafer or substrate is thoroughly cleaned to eliminate any contaminants, such as organic residues, oxides, or particles. These impurities can compromise adhesion, causing defects like delamination or voids, which can propagate through subsequent layers and degrade performance. Cleaning techniques typically involve solvent washes with acetone or isopropyl alcohol, acid treatments to remove native oxides, or plasma cleaning to eliminate stubborn particles [7]. A well-prepared surface ensures that the substrate is contamination-free and chemically active, promoting strong adhesion between the metal layer and the underlying material.

Next, activation and seed layer deposition are performed to initiate the deposition process and ensure uniform metal growth. If the surface is non-conductive or complex, it is activated by applying a thin catalytic layer, such as palladium, to promote the chemical reaction required for deposition. In the case of copper electroplating, a seed layer of copper is often deposited through PVD or electroless plating. This seed layer provides a continuous conductive path for electroplating and ensures uniform metal growth across the entire surface. Gaps or defects in the seed layer [8] can lead to non-uniform deposition and structural weaknesses, making this step crucial for achieving defect-free metal films.

Electrolyte preparation and bath control form the core of the ECD process, as the electrolyte solution provides the metal ions needed for deposition. The bath must maintain a stable concentration of metal ions, along with the proper pH and chemical balance, to avoid defects such as rough surfaces or precipitates. Additives such as brighteners, suppressors, and leveling agents are often included to improve the smoothness and grain structure of the deposited film. The temperature of the electrolyte bath is carefully regulated to optimize the deposition rate and ensure consistent results. Regular monitoring and adjustment of the electrolyte composition are essential to maintain bath stability and ensure uniform film growth across multiple wafers.

During deposition, current control and monitoring are essential to achieve precise metal film growth. Electroplating relies on applying a current that drives metal ions toward the cathode, where they are reduced and deposited. The current density directly influences the deposition rate and film uniformity. DC plating [9] applies a continuous current, but pulse plating, which alternates between plating and rest cycles, offers improved control by allowing ions to redistribute between pulses. Pulse plating is particularly effective for high-aspect-ratio features, ensuring uniform deposition in narrow trenches and deep vias [10]. In-situ monitoring systems are often employed to track film thickness in real-time, preventing over-deposition and ensuring tight process control [11].

After deposition, post-deposition cleaning and annealing are performed to remove residual chemicals and enhance the properties of the deposited film. Cleaning with deionized water and solvents eliminates any remaining electrolyte or organic residues, preventing contamination in subsequent process steps. Plasma treatments may also be used to ensure the surface is free from microscopic particles. Annealing [12], which involves heating the deposited film, helps relieve internal stress, promote grain growth, and improve film adhesion. It also enhances the electrical properties of the metal layer by reducing resistivity and eliminating any voids or microcracks that may have formed during deposition [13].

Each of these steps—surface preparation, activation, electrolyte management, current control, and post-deposition treatment—plays a critical role in ensuring the quality and reliability of electroplated metal films. Through precise control over these stages, manufacturers can achieve defect-free, uniform layers essential for modern semiconductor devices, enabling high-performance interconnects, TSVs, and advanced packaging solutions.

7.6   Materials Used in Electrochemical Deposition (ECD)

The success of ECD relies heavily on the precise selection and control of materials. These include metals, alloys, electrolytes, and additives, each contributing to the quality, reliability, and functionality of the deposited films. Carefully balancing these components ensures the creation of films with desirable properties such as high conductivity, mechanical stability, corrosion resistance, and surface uniformity, which are critical for modern semiconductor manufacturing and advanced packaging technologies.

Metals like copper, gold, nickel, and silver are commonly used in ECD processes, each chosen for its specific electrical, chemical, or mechanical properties. Copper is widely favored for forming interconnects, RDLs, and TSVs due to its excellent electrical conductivity and low resistivity. Copper's ability to resist electromigration further enhances the reliability of interconnects, especially in high-performance and high-frequency devices. Gold, with its superior corrosion resistance and biocompatibility [14], is often used for contact pads, connectors, and bonding bumps. These properties make gold indispensable for applications requiring long-term stability, such as in MEMS [15] sensors, RF components, and medical devices. Nickel plays a dual role, serving as both a barrier layer to prevent copper diffusion into surrounding dielectrics and as a wear-resistant surface finish. Silver, known for its high electrical and thermal conductivity, finds niche applications in power electronics and RF circuits, although its tendency to oxidize requires protective coatings.

In addition to pure metals, alloys like tin-silver (Sn-Ag) and nickel-phosphorus (Ni-P) offer enhanced material properties suited for specific applications [16]. Tin-silver alloys are a common replacement for lead-based solders, providing mechanical strength and thermal stability in flip-chip bonding and wafer bumping processes. The use of lead-free alloys aligns with environmental regulations while maintaining the performance needed for consumer electronics, automotive systems, and wearables. Nickel-phosphorus alloys, on the other hand, are valued for their corrosion resistance and durability, often used in barrier layers and contact surfaces to enhance reliability and prevent degradation over time. The inclusion of phosphorus in the alloy improves the mechanical properties and extends the lifespan of connectors and pads subjected to repeated thermal and mechanical stress.

The electrolytes used in ECD are equally important, as they supply the metal ions required for deposition. These electrolyte solutions must be carefully formulated to maintain stability and prevent defects in the deposited film. Sulfate-based electrolytes, such as copper sulfate, are commonly used for copper plating, providing a stable metal ion source with excellent solubility. Nitrate-based solutions, such as silver nitrate, support high-quality silver deposition by maintaining consistent ion availability throughout the process. Acids like sulfuric acid and hydrochloric acid are often added to regulate the pH of the electrolyte, ensuring smooth deposition and preventing unwanted precipitation. The precise control of these electrolyte components ensures uniform film growth, which is critical for the performance and reliability of interconnects, vias, and contact surfaces.

Additives play a crucial role in refining the properties of the deposited films, helping to control grain structure, surface roughness, and mechanical strength. Leveling agents ensure even distribution of metal across the substrate, preventing rough spots and improving the smoothness of the film. This is particularly important for high-aspect-ratio features, where maintaining uniform coverage can be challenging. Brighteners modify the microstructure of the metal, enhancing the reflectivity and appearance of the surface [17], which is valuable in decorative applications or contact areas where visibility is a factor. Suppressors slow down deposition in certain regions, helping to prevent over-plating and ensuring consistent thickness. In pulse plating processes [18], suppressors [19] enable better control of deposition, especially in narrow trenches and deep vias, where ion distribution is critical for achieving defect-free films.

The combination of high-purity metals, precisely controlled alloys, well-balanced electrolytes, and specialized additives makes ECD a versatile and powerful tool for semiconductor manufacturing. By carefully managing these materials, manufacturers can create films with optimized electrical, thermal, and mechanical properties that meet the rigorous demands of advanced applications, from 3D integrated circuits and TSVs to wearables and high-frequency devices. The ability to fine-tune these components ensures that ECD remains an indispensable technique in achieving reliable, high-performance microelectronics.

7.7   Applications of ECD in Semiconductor Manufacturing

7.7.1   Copper Interconnects: Supporting Signal Integrity in Complex Circuits

As integrated circuits evolve with chiplet architectures and heterogeneous integration, the function of copper interconnects extends beyond traditional signal routing. With the shrinking of technology nodes, surface roughness, grain structure, and interface smoothness significantly impact signal integrity. ECD enables void-free copper deposits with tightly controlled grain boundaries, minimizing the skin effect in high-frequency applications, such as 5G modules and AI processors. This allows the signal to travel through the bulk of the interconnect rather than concentrating on the surface, reducing signal loss [20].

Additionally, the continuous push toward ultra-low-power devices has made ECD's precision essential in forming multi-layered interconnect stacks with optimized thicknesses for both performance and power management. The ability to deposit copper layers of varying thickness ensures the right balance between resistance and capacitance, reducing parasitic losses in dense circuit layouts.

7.7.2   Wafer Bumping: Enabling Heterogeneous Integration Through Advanced Microbumping

ECD has transformed wafer bumping by enabling microbump [21] formation—smaller and more precisely spaced bumps that facilitate die-to-die interconnections. In AI chips and HBM, microbumped interconnects support rapid data transfer between logic and memory, minimizing latency.

The latest ECD innovations focus on co-deposition techniques that integrate solder bumps with barrier layers, eliminating the need for additional steps [22]. This not only simplifies the manufacturing process but also improves bump reliability [23], preventing common issues such as bump fatigue and cracking under thermal stress. Furthermore, lead-free bumping solutions continue to advance, aligning with global environmental standards while maintaining excellent electrical performance.

7.7.3   Redistribution Layers (RDLs): Bridging the Gap Between Chips and External Interfaces

ECD enables the fabrication of complex RDLs that enhance connectivity within FOWLP. Beyond their role in I/O routing, these layers now integrate power management circuits and passive components, such as decoupling capacitors, directly into the packaging. This shift minimizes the need for external components, reducing the overall size and cost of the package.

Another emerging trend is the use of thin copper RDLs optimized for RF circuits and antenna-in-package (AiP) modules. By reducing signal attenuation and improving thermal dissipation, ECD allows RDLs to handle the high-frequency demands of 5G and mmWave technologies. As IoT devices evolve, ECD's role in producing robust, high-performance RDLs ensures that flexibility and reliability are not compromised in compact designs.

7.7.4 Thin-Film Deposition for MEMS: Functional Coatings for Enhanced Device Reliability

In MEMS manufacturing, ECD plays a pivotal role in creating functional thin films that enhance both mechanical performance and environmental durability. ECD-deposited films of nickel or gold are used to coat moving parts in MEMS sensors, reducing friction and preventing corrosion, which is critical for automotive sensors and biomedical implants operating in harsh environments.

Recent advances in multi-layer thin-film deposition through ECD have enabled the production of gradient films, where properties such as conductivity and hardness vary across the thickness. This customization enhances the functionality of MEMS devices used in precision applications, such as optical switches, accelerometers, and energy harvesters [24]. Moreover, ECD's ability to deposit films on non-planar surfaces makes it indispensable for wearable sensors and flexible electronics, where mechanical flexibility must be balanced with electrical conductivity.

7.8   ECD-Based Solutions for Addressing Vulnerabilities

7.8.1   Mitigating Electromigration in Copper Interconnects Through Microstructural Control

Electromigration, the movement of metal atoms under high current densities, can lead to void formation and eventual failure of interconnects. ECD provides a solution by controlling grain boundaries during copper deposition, ensuring uniform grain structures that resist electromigration. Through techniques like pulse plating, manufacturers can precisely control the grain size, reducing high-density current paths that accelerate atomic migration. This enhanced grain control not only improves electrical performance but also extends the lifespan of copper interconnects, ensuring they remain reliable in high-frequency and high-power devices.

7.8.2   Achieving Void-Free Deposits with Advanced Plating Techniques

Void formation, particularly in TSVs and fine-pitch interconnects, can compromise signal integrity and mechanical stability. ECD offers precise control over the deposition process, ensuring dense, void-free metal films even within high-aspect-ratio features. Pulse-reverse electroplating [25] enhances ion distribution, eliminating pockets where voids might form. This technique ensures uniform copper filling in deep trenches and vias, critical for maintaining high-speed data transmission and minimizing signal degradation in complex, multi-layered chips.

7.8.3   Reducing Stress-Induced Cracking with Multi-layer Plating and Additives

Mechanical stress during and after metal deposition can result in cracking, delamination, and film failure, particularly in thick metal layers or complex structures. ECD enables the formation of multi-layer films where stress can be distributed across layers, reducing the risk of cracking. Additionally, the use of additives such as suppressors and grain refiners [26] allows manufacturers to control film stress and prevent mechanical failure. These additives ensure smooth deposition and improve the metal's microstructure, enhancing both ductility and toughness.

In MEMS applications or flexible electronics, where mechanical deformation is frequent, ECD-deposited films maintain their performance under bending or thermal cycling, improving both operational life and mechanical robustness.

7.8.4   Improving Adhesion and Reliability Through Seed Layers and Annealing

Poor adhesion between metal films and substrates can cause delamination, electrical failure, or void formation over time. ECD processes incorporate seed layers, such as a thin layer of copper or nickel, to ensure that the deposited metal adheres properly to the substrate [27]. Electroless seed layers further enhance adhesion by providing continuous, uniform coverage across non-planar surfaces. In addition, annealing treatments are used to relieve residual stress, improve adhesion, and promote grain growth, ensuring strong mechanical bonding and stable electrical performance throughout the life of the device.

7.8.5   Corrosion Protection with Electroplated Gold and Nickel Layers

Corrosion can compromise the long-term reliability of interconnects, connectors, and solder bumps, particularly in harsh environments such as automotive or aerospace applications. ECD offers robust corrosion protection by depositing layers of gold or nickel on the surface of sensitive components. These metals not only provide oxidation resistance but also enhance electrical conductivity, ensuring consistent performance even under exposure to humidity, chemicals, or extreme temperatures.

In specialized applications, such as medical electronics and RF devices, thin gold layers provide both biocompatibility and electrical stability, maintaining signal integrity and minimizing the risk of performance degradation over time. Nickel, on the other hand, serves as an intermediate barrier layer, preventing the migration of base metals into the surface layer and preserving contact reliability.

7.9   Challenges and Process Optimization in ECD

7.9.1   Void Formation and Poor Coverage in High-Aspect-Ratio Features

One of the biggest challenges in ECD is ensuring consistent metal deposition within high-aspect-ratio structures, such as deep trenches, TSVs and narrow RDLs. In these cases, conventional deposition techniques can result in voids, seams, or uneven coverage, leading to signal loss, increased resistance, and structural failure. In the narrow confines of deep vias, metal ions struggle to penetrate evenly, causing poor filling at the bottom while the upper sections over-deposit.

To mitigate this, advanced pulse-reverse plating is employed, where periodic current reversal pulls metal ions toward harder-to-reach regions, ensuring uniform filling. Researchers are also exploring superconformal plating chemistries—where deposition rates are accelerated at the deepest parts of the feature—using highly specialized surface-active additives. These additives selectively enhance ion movement to areas with limited access, reducing void formation and ensuring continuous deposition along the entire depth of vias and trenches.

7.9.2   Controlling Film Stress and Grain Size for Reliability

The internal stress of deposited films poses a significant threat to mechanical reliability, especially as semiconductor layers shrink to nanometer scales. Films under stress can crack, warp, or delaminate, causing failures in interconnects, TSVs, and MEMS devices. Additionally, stress impacts grain structure, influencing electrical properties like resistivity and electromigration resistance.

Process optimization involves tailoring the grain size of deposited films through precise current modulation. Fine-grained structures are more resistant to mechanical fatigue but may exhibit higher resistivity, while large-grained copper films offer better conductivity but are more prone to stress-induced cracking. To address these trade-offs, multi-layered plating techniques are used, where alternating fine- and coarse-grain layers balance electrical and mechanical performance.

Another approach is stress-engineering through additives, which modify the growth dynamics of the deposited film. Stress-relieving agents embedded in the electrolyte help absorb internal stress during deposition, preventing cracks and ensuring better mechanical stability. Additionally, annealing processes post-deposition further reduce residual stress by promoting grain coarsening and improving film adhesion.

7.9.3   Uniformity Issues Across Large Wafers: A Scaling Challenge

As wafer sizes increase—particularly with the transition to 300mm and 450mm wafers—achieving uniform film thickness becomes a challenge. Differences in current density distribution across the wafer surface can cause center-to-edge thickness variations, resulting in non-uniform electrical performance and yield loss. Ensuring film uniformity across large substrates is essential for maintaining consistent signal integrity and thermal performance in advanced devices [28].

To address these challenges, rotational systems are integrated into plating baths to dynamically adjust the position of the wafer, ensuring even current distribution. Multi-zone anode configurations are also used, where the plating bath is divided into sections, each with a carefully controlled current profile. This prevents over-deposition at the edges while ensuring the center receives adequate metal coverage. Additionally, real-time thickness monitoring systems are deployed to detect deviations and adjust plating parameters mid-process, ensuring uniformity across the entire wafer.

7.9.4   Process Optimization Techniques: Pulse Plating, Additives, and Dynamic Bath Control

Optimization of the ECD process relies on innovative plating techniques, specialized additives, and advanced bath management systems. Pulse plating—in which the current alternates between on and off cycles—allows for greater control over deposition, promoting uniform ion distribution across complex surfaces. Pulse-reverse plating further improves uniformity by preventing over-plating and filling difficult-to-reach regions in TSVs and narrow trenches.

Additives play a critical role in controlling the microstructure and stress levels of the deposited films. Suppressors limit the growth rate in specific areas to ensure even coverage, while brighteners enhance the smoothness and reflectivity of the deposited metal, critical for contact surfaces. Grain refiners control the crystal structure of the metal, promoting smaller, more uniform grains that improve mechanical properties and resistance to electromigration.

Bath control has evolved with the integration of smart monitoring systems, where sensors continuously track pH levels, temperature, ion concentration, and additive depletion rates. These systems provide real-time feedback, automatically adjusting parameters to maintain optimal plating conditions [29]. AI-powered bath management tools are emerging, capable of predicting when maintenance or replenishment is needed, reducing downtime and improving process yield.

7.10 Emerging Trends and Innovations in ECD

7.10.1 Pulse-Reverse Electroplating: Achieving Superior Uniformity

Pulse-reverse electroplating represents a significant innovation in metal deposition, addressing one of the key challenges in traditional ECD is the uniformity across complex surfaces. This technique alternates between forward and reverse currents, redistributing metal ions more evenly across the substrate. By reversing the current periodically, pulse-reverse plating mitigates the accumulation of excessive material in certain areas, ensuring better coverage in high-aspect-ratio structures, such as deep vias or narrow trenches.

This method also improves surface smoothness and grain structure control, enhancing both mechanical and electrical properties. Pulse-reverse plating is increasingly used in 3D ICs and TSV filling, where uniform copper deposition is essential for maintaining signal integrity. In addition, it reduces defects like seams, voids, and over-plating, extending the lifespan of interconnects by minimizing the risks associated with current crowding and electromigration.

7.10.2 Hybrid ECD Processes: Combining Electroplating with Electroless Deposition

Hybrid ECD processes, which integrate electroplating and electroless deposition, offer enhanced flexibility in fabricating advanced microelectronic structures. Electroplating provides fast, scalable deposition for conductive surfaces, while electroless deposition excels at covering non-conductive areas or irregular geometries where electric fields cannot penetrate effectively. By combining these methods, manufacturers can achieve seamless metal coatings across both planar and complex surfaces, ensuring uniform film growth and enhanced adhesion.

For example, seed layers deposited through electroless plating can prime the substrate for subsequent electroplating, ensuring continuous metal growth without gaps or discontinuities. Hybrid ECD processes are particularly valuable in heterogeneous integration, where different materials and surfaces must be interconnected seamlessly. They also support the creation of multi-functional coatings, such as those used in MEMS sensors or wearable electronics, where both conductive properties and environmental resistance are required.

7.10.3 Precision Mask Jet ECD Process Flow

The mask jet ECD process is illustrated in Fig. 7.3. is an innovative approach that merges traditional ECD techniques with precision masking technology to enable highly selective metal deposition. The process begins by applying a temporary mask to the substrate using a mask jet system, which sprays or prints the mask material in a precise pattern. This masking

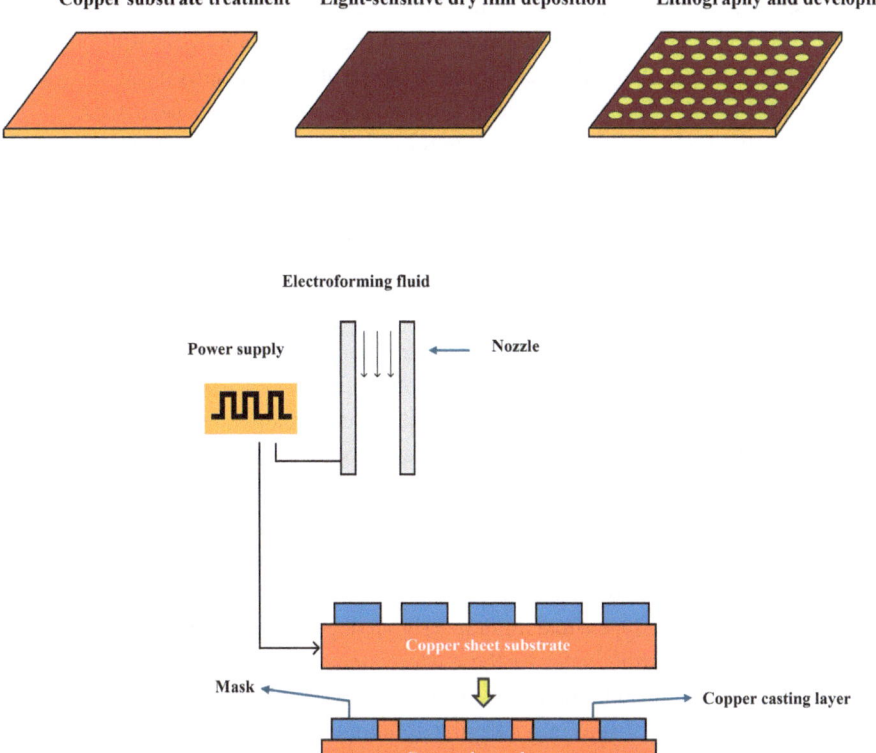

Fig. 7.3 Process flow diagram for mask jet electrochemical deposition (ECD)

step defines areas where metal will be selectively deposited, ensuring that only specified regions receive metal coverage while other areas remain untouched. The use of masks in this way minimizes material waste, streamlines the deposition process, and significantly improves the precision of feature formation.

After the masking step, the substrate enters the electroplating phase, where a metal—such as copper, nickel, or gold—is deposited only within the exposed regions. The electroplating solution facilitates a controlled redox reaction that drives metal ions toward the masked substrate, filling high-aspect-ratio features like vias, trenches, or intricate microstructures. This step is critical for creating robust, conductive pathways in applications that demand fine structural resolution, such as interconnects in semiconductor packaging and MEMS.

Once the metal deposition reaches the desired thickness, the mask is removed, revealing a clean and accurately patterned metal layer. This final step is often followed by a rinsing process to remove any residual contaminants, ensuring high-quality, defect-free metal features. The mask jet ECD process is particularly advantageous for advanced electronic applications, where precision, material efficiency, and structural integrity are essential. By incorporating established ECD technology with this innovative masking approach, mask jet ECD provides a versatile and cost-effective solution for modern fabrication challenges, enabling high-performance interconnects and advanced microstructures in next-generation devices.

7.10.4 ECD for Advanced Interconnects in AI and 5G Applications

The rise of AI accelerators, 5G networks, and high-bandwidth memory (HBM) is driving the need for high-density interconnects capable of supporting ultra-fast data transmission with minimal power loss. ECD is playing a crucial role in enabling these next-generation devices by facilitating the deposition of copper interconnects with superior conductivity and electromigration resistance.

In AI chips, ECD enables the formation of intricate multi-layer interconnect stacks that ensure low-latency data transfer between processing cores and memory modules. Similarly, in 5G antennas and RF components, ECD supports the fabrication of high-frequency interconnects and redistribution layers that minimize signal degradation across wide bandwidths. Pulse plating techniques further optimize these interconnects by preventing defects, ensuring that devices meet the demanding requirements of low-power, high-frequency applications.

ECD also plays a pivotal role in packaging solutions for AI and 5G systems, such as FOWLP and antenna-in-package (AiP) modules. These applications require interconnects with precise thickness control, low resistance, and the ability to withstand thermal stress—all of which are achieved through advanced ECD processes.

7.10.5 Sustainable ECD Processes: Green Chemistry and Recycling of Electrolytes

With growing concerns about environmental sustainability, the semiconductor industry is focusing on developing green ECD processes that reduce waste and chemical consumption. Traditional ECD involves the use of toxic chemicals and produces electrolyte waste that requires costly disposal. In response, researchers are developing green chemistries that replace hazardous chemicals with environmentally friendly alternatives.

Recycling and reusing electrolytes have become key strategies in sustainable ECD practices. Advanced filtration and ion-exchange systems allow manufacturers to recover valuable metal ions from used baths, reducing the need for fresh electrolytes. In some cases, closed-loop systems are implemented, where the same electrolyte solution is continuously monitored, purified, and reused, minimizing waste and operational costs.

Another promising approach involves the use of low-energy ECD systems that operate at lower temperatures and consume less power. These systems are designed to reduce the overall carbon footprint of the process while maintaining high-quality metal deposition. Sustainable ECD processes align with the semiconductor industry's carbon-neutral goals and circular economy initiatives, ensuring that future technologies are both high-performance and environmentally responsible.

7.11   Industry Vendors and Ecosystem

The ECD ecosystem involves a dynamic network of vendors, suppliers, research institutions, and manufacturers collaborating to advance microelectronics manufacturing. This ecosystem plays a crucial role in enabling high-performance interconnects, packaging solutions, and advanced technologies such as AI and 5G. Vendors specializing in ECD materials, equipment, and process optimization tools are essential partners in scaling up next-generation packaging solutions, while collaborative efforts with academic institutions and research centers continue to push the boundaries of what is possible. Below is a closer look at the key players and market dynamics shaping the future of ECD.

7.11.1   Key Players in the ECD Ecosystem

Several companies have emerged as global leaders in ECD technology, providing chemistries, process tools, and solutions to support the semiconductor industry. These companies supply essential materials like electrolytes, additives, and plating equipment used across various stages of chip manufacturing and packaging.

Atotech: Atotech is a global leader in electroplating chemicals and systems, offering solutions tailored for semiconductor packaging, interconnects, and MEMS fabrication. The company provides advanced chemistries that enhance bath stability, reduce voids, and enable high-aspect-ratio plating. Atotech's focus on environmentally friendly products reflects the industry's shift toward sustainable processes.

Technic: Technic specializes in the development of plating solutions, equipment, and electrolytes designed for precision applications. It offers tailored ECD processes for semiconductor interconnects, including copper, gold, and nickel electroplating. Technic's emphasis on process customization makes it a trusted partner for companies developing high-frequency RF devices and advanced memory modules.

Dow: A major player in chemical technologies, Dow provides specialized additives and electrolytes for ECD. Its innovations focus on improving film smoothness, stress control, and electromigration resistance. Dow's products are widely used in TSVs, fan-out packaging, and thin-film coatings for MEMS devices.

Entegris: Entegris plays a critical role in the ECD ecosystem by offering chemical management solutions and high-purity materials essential for defect-free plating. Its advanced filtration and purification systems ensure that plating baths remain stable, helping manufacturers achieve uniform films with reduced defect rates. Entegris also supports closed-loop chemical recycling systems, aligning with the industry's move toward sustainable manufacturing.

7.11.2 Global Market Trends: Adoption in Advanced Packaging and Heterogeneous Integration

The global market for ECD is experiencing significant growth, driven by demand for advanced packaging solutions and heterogeneous integration. The shift from traditional packaging to 3D ICs, FOWLP, and SiP technologies has increased the need for reliable ECD processes. As devices shrink and interconnect density increases, ECD offers the precision required for forming high-performance copper interconnects, microbumps, and TSVs.

The adoption of ECD processes has also accelerated with the rise of AI, 5G, and automotive electronics. These technologies require high-frequency interconnects with minimal power loss, which ECD supports through pulse plating techniques and advanced additives. Additionally, the transition to lead-free packaging solutions and the integration of power management components within RDLs have further fueled the demand for ECD-based processes.

Regional growth patterns reflect the increasing adoption of heterogeneous integration and advanced packaging. The Asia-Pacific region—home to major semiconductor manufacturers such as TSMC, Samsung, and ASE Group—continues to dominate the ECD market. However, with government initiatives like the CHIPS and Science Act in the U.S. and European investments in semiconductor ecosystems, North America and Europe are witnessing renewed growth in ECD adoption.

Conclusion:
This chapter explored the integral role of electrochemical deposition (ECD) in modern semiconductor manufacturing and advanced packaging. ECD continues to evolve, offering precise metal deposition solutions that address the complex needs of high-performance interconnects, TSVs, microbumps, RDLs, and MEMS devices. Innovations such as pulse-reverse plating, hybrid processes, and sustainable chemistry have pushed the boundaries of what is achievable, enabling AI, 5G, automotive electronics, and wearable technologies to thrive. The growing involvement of key vendors

like Atotech, Technic, Dow, and Entegris, along with collaborative efforts within the ecosystem, ensures that ECD remains scalable, reliable, and sustainable. As devices continue to shrink and integrate more functionality, ECD will play a pivotal role in delivering uniform, defect-free metal layers critical for future breakthroughs in semiconductor technologies.

References

1. Paunovic, M. *Fundamentals of Electrochemical Deposition*. John Wiley & Sons, 2006.
2. Schlesinger, Mordechay, and Milan Paunovic, eds. *Modern Electroplating*. John Wiley & Sons, 2011.
3. Djokić, Stojan S., and Pietro L. Cavallotti. "Electroless deposition: theory and applications." In *Electrodeposition: Theory and Practice* (2010): 251–289.
4. Alesker, Maria, et al. "Palladium/nickel bifunctional electrocatalyst for hydrogen oxidation reaction in alkaline membrane fuel cell." *Journal of Power Sources* 304 (2016): 332–339.
5. Elendu, Oyidia, et al. "Use of a mixed formaldehyde and sodium hypophosphite reducing agent bath in the electroless synthesis of Cu-Ni-Mo-P electro-catalyst active for glycerol oxidation." *International Journal of Electrochemical Science* 10.12 (2015): 10792–10805.
6. Gamburg, Yuliy D., et al. "Technologies for the electrodeposition of metals and alloys: electrolytes and processes." In *Theory and Practice of Metal Electrodeposition* (2011): 265–316.
7. "Circuits, Integrated." *Operating Procedures*, 2012.
8. Woo, Tae-Gyu, Il-Song Park, and Kyeong-Won Seol. "Effects of various metal seed layers on the surface morphology and structural composition of the electroplated copper layer." *Metals and Materials International* 15 (2009): 293–297.
9. Lee, Wun-Hsing, Sen-Cheh Tang, and Kung-Cheng Chung. "Effects of direct current and pulse-plating on the co-deposition of nickel and nanometer diamond powder." *Surface and Coatings Technology* 120 (1999): 607–611.
10. Dixit, Pradeep, and Jianmin Miao. "High aspect ratio vertical through-vias for 3D MEMS packaging applications by optimized three-step deep RIE." *Journal of the Electrochemical Society* 155.2 (2007): H85.
11. Muramatsu, H., et al. "In-situ monitoring of microrheology on electrochemical deposition using an advanced quartz crystal analyzer and its application to polypyrrole deposition." *Journal of Electroanalytical Chemistry* 322.1-2 (1992): 311–323.
12. Van Laarhoven, Peter J.M., et al. *Simulated Annealing*. Springer Netherlands, 1987.
13. Weitz, Stefan, André Clausner, and Ehrenfried Zschech. "Microcracking in On-Chip Interconnect Stacks: FEM Simulation and Concept for Fatigue Test." *Journal of Electronic Materials* (2024): 1–9.
14. Konwar, Gargi, Sachin Rahi, and Shree Prakash Tiwari. "Exploration of a Cellulose-Based Biocompatible Gate Dielectric for Low-Voltage Organic Transistors." *IEEE Journal on Flexible Electronics* 2.5 (2023): 383–389.
15. Kottur, Himanandhan Reddy, et al. "Enhancing the MEMS Gyroscope Physical Assurance Using Quantum Sensing." In *2024 IEEE Research and Applications of Photonics in Defense Conference (RAPID)*. IEEE, 2024.

16. Bandayagari, Jyothsna. *Effects of Adding Nickel and Bismuth in SAC Solder for Second Level Interconnects Under Aging and Multiple Reflows*. MS thesis, State University of New York at Binghamton, 2024.
17. Xu, Yanqiu, et al. "Principles and Applications of Electrochemical Polishing." *Journal of The Electrochemical Society* 171.9 (2024): 093506.
18. Hilton-Tapp, H., J. Kelly, and D. Weston. "Designing effective plating baths for use in the pulse-reverse plating of copper nanocomposite coatings." *Transactions of the IMF* 101.4 (2023): 179–188.
19. Su, You-Jhen, et al. "Effects of suppressors on the incorporation of impurities and microstructural evolution of electrodeposited Cu solder joints." *Journal of the Taiwan Institute of Chemical Engineers* 149 (2023): 104956.
20. Borges, Juliano, et al. "Development of a Plasma Etching Process of Copper for the Microfabrication of High-Density Interconnects in Advanced Packaging." In *2023 IEEE 73rd Electronic Components and Technology Conference (ECTC)*. IEEE, 2023.
21. Datta, Madhav. "Manufacturing processes for fabrication of flip-chip micro-bumps used in microelectronic packaging: An overview." *Journal of Micromanufacturing* 3.1 (2020): 69–83.
22. Taheri-Ledari, Reza, et al. "A brief survey of principles of co-deposition method as a convenient procedure for preparation of metallic nanomaterials." *Journal of Alloys and Compounds* (2024): 173509.
23. Yan, Lei, et al. "Reliability Analysis of Flip-Chip Packaging GaN Chip with Nano-Silver Solder BUMP." *Micromachines* 14.6 (2023): 1245.
24. Hossain, Md Imran, et al. "MEMS-based energy harvesting devices for low-power applications—a review." *Results in Engineering* (2023): 101264.
25. Baiocco, Gabriele, et al. "Study on pulse-reverse electroplating process for the manufacturing of a graphene-based coating." *Materials* 16.2 (2023): 854.
26. Liu, Jin-Hao, et al. "Influence of macroscale dimension on the electrocrystallization of Cu pad and redistributed layer in advanced packaging." *Advanced Engineering Materials* (2024): 2400304.
27. Wang, Zhao, et al. "Enhancing Adhesion and Reducing Ohmic Contact through Nickel—Silicon Alloy Seed Layer in Electroplating Ni/Cu/Ag." *Materials* 17.11 (2024): 2610.
28. Zhang, Chi, et al. "Study on the 12 in. wafer uniformity of high aspect ratio TSV filling by using rotation cathode." *Microelectronic Engineering* 292 (2024): 112181.
29. Battiato, Sergio, et al. "Composition-controlled chemical bath deposition of Fe-doped NiO microflowers for boosting oxygen evolution reaction." *International Journal of Hydrogen Energy* 48.48 (2023): 18291–18300.

Testing and Reliability in Advanced Packaging   8

8.1   Introduction to Testing and Reliability in Packaging

Testing and reliability [1] are essential pillars of advanced packaging, ensuring that devices consistently perform as intended across a variety of operating conditions. With the growing complexity of IC packages, which now integrate MCMs, 3D architectures, a thorough and evolving testing framework is required. These frameworks validate the mechanical, thermal [2], electrical, and chemical resilience of each package throughout its lifecycle.

Reliability is more than just a technical requirement; it directly influences product longevity, user safety, and market reputation. Even seemingly minor failures–such as microcracks, delamination [3], or defective interconnects—can trigger catastrophic system failures or degrade performance over time, leading to expensive recalls and loss of consumer trust. The impact is especially significant in industries such as automotive, aerospace, telecommunications, and healthcare, where device reliability is non-negotiable.

Key challenges in ensuring packaging reliability include thermal stress caused by temperature fluctuations, mechanical fatigue from vibrations or shocks, and corrosion resulting from exposure to moisture or harsh environments. Modern testing processes are designed to identify these vulnerabilities at every stage of development, from design validation to production and field deployment. To address potential failure points, manufacturers employ a multi-disciplinary [4] approach, combining environmental simulations, accelerated stress testing, and in-situ monitoring to ensure both performance stability and structural integrity over extended operational lifetimes.

Incorporating proactive reliability testing early in the design phase has become crucial, as packaging technologies now feature denser, heterogeneous components with tighter interconnects. As a result, ensuring that these systems maintain integrity throughout their operational environments—which may involve exposure to high currents, varying temperatures, and mechanical shocks—requires advanced testing techniques that evolve alongside new packaging innovations [5].

8.2   Types of Packaging Reliability Tests

Reliability testing for advanced packaging involves evaluating the ability of electronic components to function under mechanical, environmental, electrical, and chemical stresses throughout their lifecycle. These tests provide insights into potential vulnerabilities, helping manufacturers improve product designs and ensure that devices meet performance and safety standards across industries like automotive, aerospace, telecommunications, and consumer electronics. Below are the primary types of packaging reliability tests and their specific roles in ensuring robust systems (Fig. 8.1).

8.2.1   Mechanical Testing

Mechanical testing is crucial to determine the robustness of advanced packaging systems against physical stress during transportation, assembly, and usage. As devices become more compact and integrated, even small vibrations or shocks can lead to fatigue in solder joints or microbumps [6], ultimately compromising the electrical and mechanical integrity of the system.

Vibration testing [7] simulates the oscillations experienced during transport or operational use. This type of testing ensures that the package can withstand constant vibration without developing cracks, loosened connections, or fatigue-related failures. MCMs in automotive applications, for instance, encounter continuous vibrations from engine operations, making vibration testing essential to validate long-term reliability.

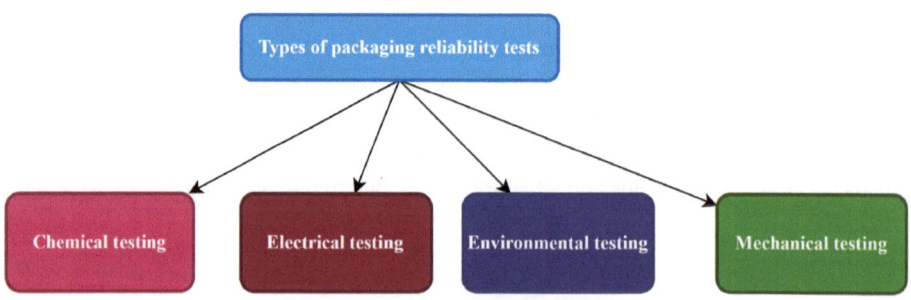

Fig. 8.1 Various reliability tests for semiconductor packaging

Shock testing [8] evaluates the impact resistance of a package by subjecting it to sudden and intense mechanical forces. Shocks may occur during handling, installation, or usage, such as when electronics are dropped or exposed to physical collisions. Shock testing ensures that interconnects, solder joints, and microbumps remain intact under these conditions. High-performance AI accelerators often undergo shock testing to verify that their chip-to-substrate connections can endure the stress of deployment without failure.

Drop testing is particularly relevant for consumer electronics and wearables, simulating accidental drops from varying heights. The goal is to verify that the package can maintain both functional and structural integrity after impact. As smartphones, smartwatches, and tablets are prone to accidental drops, rigorous drop testing ensures that these products continue to operate reliably in real-world scenarios.

8.2.2 Environmental Testing

Environmental testing [9] plays a pivotal role in assessing the durability of packaging systems when exposed to a variety of environmental stressors, such as temperature fluctuations, humidity, and corrosive atmospheres. Devices deployed in outdoor, automotive, or aerospace environments must maintain stable performance despite harsh conditions, and environmental tests provide early insights into potential failure modes.

Thermal cycling [10] and thermal shock testing simulate rapid temperature changes to detect warpage, delamination, or microcracking within the package. In 3D-stacked ICs, materials with different coefficients of thermal expansion may expand and contract unevenly, leading to internal stress. Thermal cycling exposes the package to alternating high and low temperatures over multiple cycles, identifying whether these expansion mismatches compromise the integrity of solder joints or bonding layers. In contrast, thermal shock testing rapidly transitions the package between extreme temperatures, helping to uncover instantaneous material failures.

Humidity testing [10] evaluates the package's resistance to moisture ingress. Prolonged exposure to humidity can degrade adhesives, corrode interconnects, and cause electrical shorts, especially in devices with sensitive electronic components. This type of testing is particularly important for medical devices and automotive sensors that must perform reliably in high-humidity conditions. Packages designed for outdoor environments may also undergo extended humidity tests to ensure consistent operation over the product's lifetime.

Salt fog and corrosion testing simulates the effects of corrosive environments, such as those encountered in marine applications or industrial settings. Metallic leads, connectors, and exposed surfaces are vulnerable to corrosion, which can compromise electrical performance. This test involves exposing the package to salt spray atmospheres to validate that it resists corrosion, ensuring that the package maintains both structural integrity and functionality in harsh conditions.

8.2.3   Electrical Testing

Electrical testing ensures that advanced packages meet performance standards by validating insulation, high-voltage tolerance, and signal integrity. These tests are crucial for devices where high-speed communication, power delivery, and safety are priorities, such as in telecommunications and AI systems.

The dielectric withstand test, commonly known as hipot testing, evaluates whether the package can withstand high-voltage stress without breakdown. This test confirms that the insulation between conductors is sufficient to prevent electrical shorts or leakage currents under operational voltage. High-power modules used in electric vehicles (EVs) and industrial applications often undergo hipot testing to validate their safety and reliability at elevated voltages.

The insulation resistance test measures the resistance between conductors to ensure that there is no unintended current leakage. A high insulation resistance indicates that the device can maintain its performance under typical operating conditions without experiencing electrical drift or failure. This test is especially critical in aerospace and healthcare electronics, where any leakage current can cause safety hazards or malfunctions.

Signal integrity testing evaluates the ability of high-speed signals to travel through the package without distortion, jitter, or latency. With the increasing use of 5G and AI accelerators, ensuring proper signal transmission across densely packed interconnects is essential. Signal integrity tests analyze timing errors, impedance mismatches, and crosstalk, ensuring that communication channels within the package function reliably at high data rates.

8.2.4   Chemical Testing

Chemical testing [11] ensures that the materials used in advanced packaging are resistant to chemical degradation and contamination, both of which can impair long-term performance. Packaging systems often encounter harsh chemicals during manufacturing, cleaning, or deployment, and they must resist deterioration to ensure durability.

Outgassing [12] and contamination testing identifies volatile organic compounds (VOCs) released from package materials, which could contaminate sensitive components such as optical sensors. This test is particularly important in aerospace applications, where contaminants released in a vacuum environment could interfere with precise instrumentation or cause electrical malfunctions. By detecting and controlling outgassing, manufacturers can ensure that packages remain stable and free from contamination.

Chemical resistance testing evaluates how well packaging materials withstand exposure to solvents, cleaning agents, and industrial chemicals. In environments such as medical facilities and factories, devices often require regular cleaning or sterilization, exposing them to aggressive chemicals. Chemical resistance testing ensures that materials do not crack, degrade, or lose their insulating properties when subjected to such treatments. This testing is especially relevant for wearable medical devices and industrial sensors, where frequent maintenance is required.

8.3   Common Reliability Issues in Packaging

Advanced packaging systems are increasingly complex, incorporating MCMs, 3D-stacked ICs and flexible electronics. While these innovations enhance performance and enable new functionalities, they introduce several reliability challenges. Mechanical, thermal, electrical, and environmental stresses can accumulate over time, leading to packaging failures. Understanding these common failure modes and employing targeted detection methods ensures the stability and longevity of these systems. Below are the key reliability issues encountered in advanced packaging and the methods used to identify them (Fig. 8.2).

8.3.1   Cracking, Delamination, and Warping

Cracking, delamination, and warping are common in packaging materials subjected to repeated thermal cycling or operational heat stress. Packages are composed of multiple materials—such as polymers, metals, and ceramics—each with a CTE. When exposed to varying temperatures, these materials expand and contract at different rates, causing internal stress to accumulate at their interfaces. Over time, this leads to microcracks, delamination between bonded layers, or warping of substrates. These issues degrade the mechanical integrity of the package, resulting in signal loss, poor electrical connections, or total device failure [13].

Cracking typically occurs in solder joints and microbump interconnects, weakening the pathways that connect dies or substrates. Delamination—the separation of bonded layers— is particularly problematic in 3D-stacked ICs and RDLs, where it can cause open circuits or increased resistance. Warping, often seen in FOWLPs, leads to misalignment during assembly, affecting downstream processes and device yield.

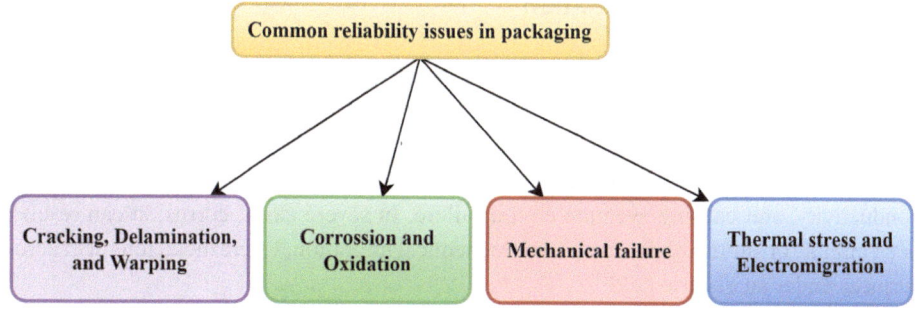

Fig. 8.2 Classifications of reliability issues in semiconductor packaging

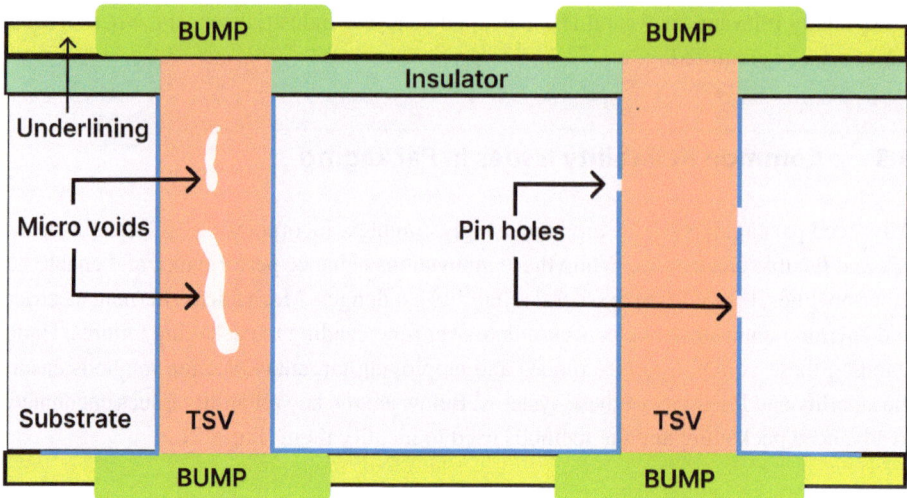

Fig. 8.3 Illustration of void formation in advanced semiconductor packaging

How It Is Detected:

Acoustic Microscopy provides high-resolution images of internal interfaces, detecting delamination and cracking by analyzing acoustic reflections from material boundaries. It is particularly useful for identifying hidden defects in multi-layer packages without disassembling them.

Thermal Cycling Tests with Real-Time Optical Profilometry track the dimensional stability of substrates and layers during temperature transitions. Any warping or deformation detected during these tests indicates a potential reliability risk.

X-Ray [14] and CT Scanning are used to visualize internal cracks or voids as depicted in Fig. 8.3, vias, and microbumps, providing non-invasive insights into the structural integrity of the package.

8.3.2 Corrosion and Oxidation

Corrosion and oxidation are significant threats to metallic interconnects, bonding pads, and exposed connectors. These issues typically arise when moisture or chemical contaminants penetrate the package. Prolonged exposure to high humidity or corrosive environments leads to oxidative reactions on metal surfaces, increasing resistance, disrupting electrical conductivity, and causing eventual circuit failure. In severe cases, corrosion can result in open circuits or short circuits due to electrochemical migration, where metal ions drift across surfaces under an electric field.

Packages deployed in outdoor environments, automotive systems, or industrial equipment are particularly susceptible to corrosion. For instance, marine electronics encounter salt-laden air that accelerates corrosion, while automotive modules experience high humidity and rapid temperature changes, creating conditions favorable to oxidation.

How It Is Detected:

Humidity Testing exposes packages to controlled moisture levels, monitoring electrical resistance over time to identify any increase caused by corrosion-induced degradation of metal surfaces.

Salt Fog Testing simulates exposure to saline environments, providing insights into the long-term durability of connectors and metal leads. Resistance measurements taken throughout the test reveal early-stage corrosion.

Four-Point Probe Resistance Measurements are used to detect subtle changes in the electrical resistance of interconnects, indicating the formation of oxidation layers even before they are visible.

8.3.3 Thermal Stress and Electromigration

Thermal stress and electromigration are critical reliability issues, especially in high-performance packages such as AI accelerators, power modules, and high-frequency RF devices. As these systems generate significant heat during operation, localized hotspots [15] can form, causing thermal expansion mismatches that degrade the package over time. Fatigue in solder joints, delamination of layers, and material deformation are common results of unmanaged thermal stress. If not addressed, these failures compromise signal integrity and power delivery, leading to performance degradation.

Electromigration [16] occurs when metal ions migrate under the influence of high current densities, creating voids in interconnects that disrupt signal paths. Over time, these voids grow, resulting in increased resistance, open circuits, or short circuits. Electromigration is particularly problematic in fine-pitch microbump interconnects used in 3D-stacked ICs, where the small cross-sectional area exacerbates the migration process (Fig. 8.4).

How It Is Detected:

Infrared Thermography captures temperature gradients across the package, identifying hotspots and areas prone to thermal fatigue. These thermal anomalies often indicate regions where solder joints or materials are likely to fail.

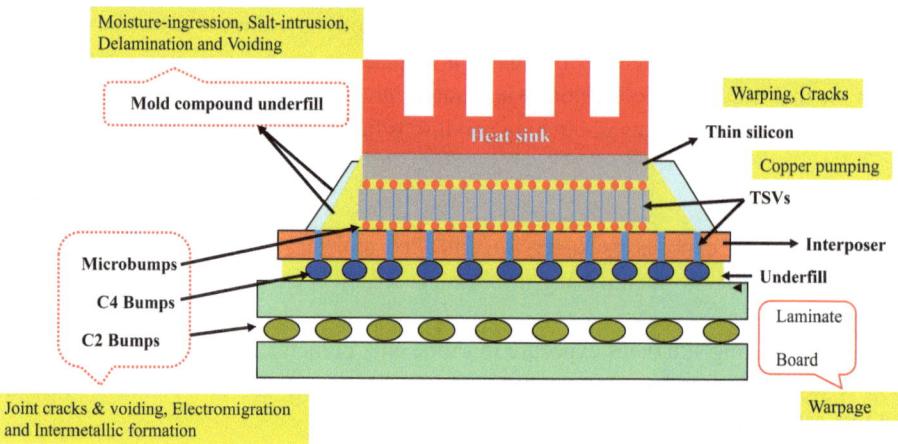

Fig. 8.4 Common reliability issues in advanced packaging

Electromigration Tests subject interconnects to constant high currents and track resistance over time. Increasing resistance signals the formation of voids or weak points, helping predict the lifespan of interconnects under operational loads.

Thermal Shock and Cycling Tests combined with time-domain reflectometry (TDR) [17] monitor the structural stability of interconnects under rapid temperature transitions, detecting failures caused by thermal expansion mismatches.

8.3.4   Mechanical Failures

Mechanical failures result from vibrations, shocks, and drops experienced during transport, assembly, or regular use. These forces can cause microcracks in solder joints, loosening of connectors, or delamination in thin-film substrates. Devices such as smartphones, automotive sensors, and aerospace modules are frequently subjected to physical stresses that challenge the durability of their packaging. Even minor mechanical disruptions can trigger intermittent faults, leading to degraded performance or total device failure over time.

Vibration-induced fatigue weakens solder joints and interconnects, while shocks during handling or accidental drops can damage fragile components within the package. For example, in surface-mount devices (SMDs), repeated shocks can cause detachment of chips from the substrate, interrupting functionality. In automotive systems, continuous vibrations from engine operations create conditions that accelerate fatigue in critical interconnects.

How It Is Detected:

Vibration Testing with Electrical Continuity Monitoring subjects packages to controlled oscillations while tracking real-time signal integrity. Any sudden drops in signal strength indicate mechanical fatigue or damaged interconnects.

Drop Testing with High-Speed Imaging analyzes the behavior of components during impact, identifying displacement or fractures caused by sudden forces. This test ensures that consumer electronics maintain structural integrity under accidental drops.

Acoustic Microscopy detects microcracks and delamination caused by mechanical stress, providing non-invasive insights into internal structures. This technique helps evaluate whether mechanical stress during transport has compromised the package.

8.4   Non-destructive Testing (NDT) Methods

Non-destructive testing methods allow manufacturers to inspect internal structures and materials without altering or damaging the package, making them suitable for use throughout the production and validation processes. These methods enable the detection of early-stage defects, misalignments, and voids, ensuring high-quality outputs without compromising yield or integrity.

X-ray and CT scanning are highly effective in visualizing internal defects in complex multi-layer packages, such as 3D-stacked ICs and FOWLPs. X-ray technology offers detailed two-dimensional imaging that identifies voids within solder joints or misaligned interconnects, and a few critical features in GPU are mentioned in Figs. 8.6 and 8.7. For more comprehensive analysis, CT scanning provides three-dimensional views, allowing engineers to inspect fine-pitch interconnects and detect defects within deep structures like TSVs. These imaging techniques are invaluable during quality control and failure diagnostics, ensuring consistent structural integrity across production batches.

Acoustic microscopy uses ultrasonic waves to detect delamination, internal cracks, and voids within multi-layer components and some examples are mentioned in the Fig. 8.5. As ultrasonic waves pass through different materials, they reflect differently depending on the material's properties. Any change in acoustic impedance reveals hidden defects that might compromise bonding quality. This method is particularly useful for inspecting encapsulated components, RDLs, and solder bumps in 3D packages, where surface-level inspections are insufficient.

Infrared thermography is another powerful tool that helps engineers identify thermal imbalances, hotspots, and areas prone to fatigue. By capturing the temperature distribution on the package surface, it reveals localized heat buildup that could lead to thermal degradation, delamination, or mechanical failure over time. Infrared thermography is widely used in testing AI accelerators, power modules, and high-performance processors, where even slight thermal anomalies can impact device reliability and performance.

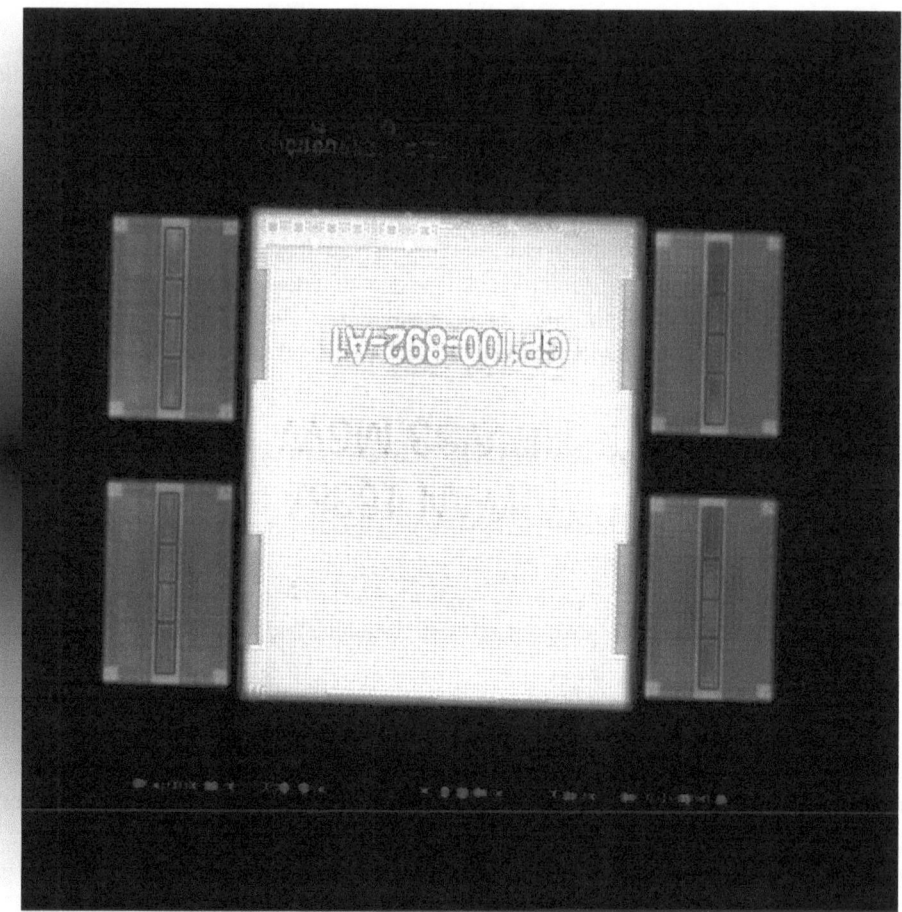

Fig. 8.5 Acoustic C-Scan analysis of gpu structure. (*Figure source* Varshney, N., Ghosh, S., Craig, P., Kottur, H. R., Dalir, H., & Asadizanjani, N. Challenges and opportunities in non-destructive characterization of stacked IC packaging: Insights from SAM and 3D X-ray analysis)

8.5   Destructive Testing Methods

Destructive testing methods involve physically altering or dismantling packages to examine internal materials, layers, and interconnects for defects. While these techniques are not reversible, they provide essential insights into failure mechanisms and material properties that non-destructive methods cannot fully uncover. These methods are particularly useful in failure analysis and material characterization, allowing manufacturers to improve designs and prevent similar issues in future production.

Cross-sectioning and failure analysis involve cutting through the package to inspect internal layers, solder joints, and TSVs. This method exposes structural defects such as cracks,

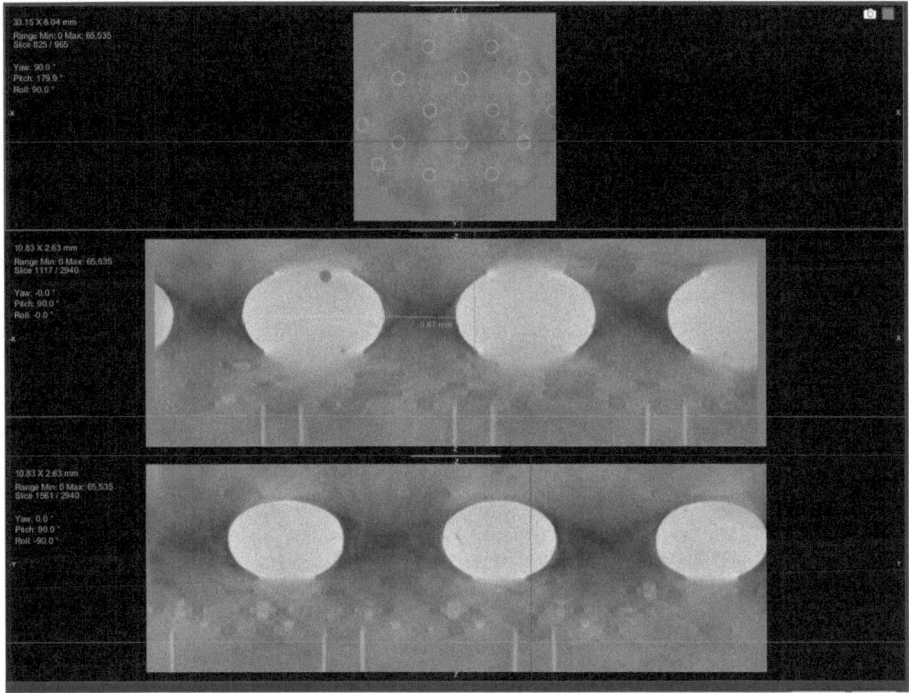

Fig. 8.6 Key features in cross-sectional 3D X-ray imaging of a GPU. (*Figure source* Varshney, N., Ghosh, S., Craig, P., Kottur, H. R., Dalir, H., & Asadizanjani, N. Challenges and opportunities in non-destructive characterization of stacked IC packaging: Insights from SAM and 3D X-ray analysis)

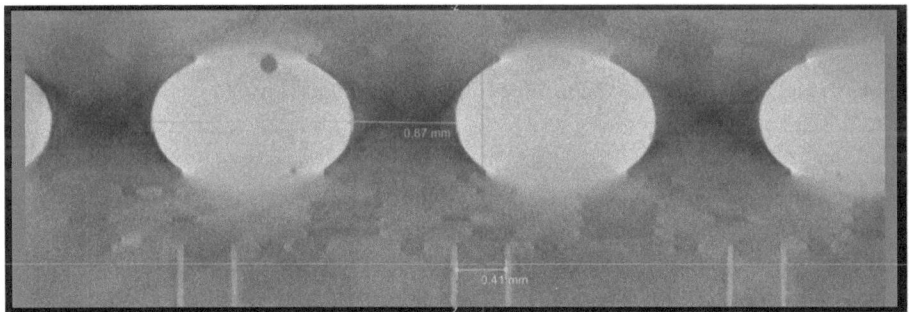

Fig. 8.7 Key features in cross-sectional 3D X-ray imaging of a GPU. (*Figure source* Varshney, N., Ghosh, S., Craig, P., Kottur, H. R., Dalir, H., & Asadizanjani, N. Challenges and opportunities in non-destructive characterization of stacked IC packaging: Insights from SAM and 3D X-ray analysis)

voids, or weak bonding points that could not be detected through external inspection. After sectioning, engineers use scanning electron microscopy (SEM) to perform detailed imaging of the cross-sections, revealing microscopic defects with high precision. This process is crucial for root cause analysis following failures detected during reliability testing or field deployment [18].

Chemical etching and layer peeling sequentially remove layers of the package to assess material composition and adhesion quality. Chemical solutions dissolve specific materials, allowing engineers to analyze the remaining layers for delamination, adhesion defects, or material inconsistencies. This method is particularly important in evaluating 3D packaging technologies, where the bonding between layers is critical for maintaining signal integrity and structural stability. Layer peeling also helps validate the compatibility of adhesives and encapsulants [19] used in packaging.

Fatigue and tensile testing measure the mechanical strength and durability of materials under stress. Fatigue testing involves applying repeated mechanical loads to the package to simulate real-world conditions, such as the vibrations experienced in automotive or aerospace applications. This test helps identify weak points in solder joints, interconnects, and bonding interfaces. Tensile testing evaluates the ultimate strength of materials by pulling them until they fracture, revealing how well they withstand tension.

These mechanical tests provide essential data for design optimization, ensuring that packages meet the required durability standards for demanding environments.

8.6   Standards and Guidelines for Reliability Testing

Reliability testing in advanced packaging is governed by a set of well-established standards and guidelines designed to ensure that electronic components meet performance, safety, and durability requirements. These standards help manufacturers evaluate packaging systems under controlled conditions, ensuring they perform reliably in their intended applications. Compliance with these standards is essential not only for product quality but also to meet regulatory requirements and customer expectations. Below is a detailed discussion of key standards and guidelines that shape reliability testing in the industry.

8.6.1   JEDEC Standards

The JEDEC Solid State Technology Association establishes industry-wide protocols for thermal management, stress testing, and material qualification. These standards are essential for ensuring that packaging materials and designs maintain stability under thermal and mechanical stress. JEDEC guidelines, such as JESD22-A104 for thermal cycling and JESD22-B102 for solderability testing, provide structured methods to assess the long-term reliability of IC packages under varying environmental conditions.

JEDEC standards also address electromigration testing and the behavior of interconnects under high current densities, which is critical for advanced systems such as AI accelerators and power modules. Following these protocols ensures that MCMs, stacked dies, and FOWLPs maintain their performance across various applications.

8.6.2   MIL-STD

MIL-STD (Military Standards) are used primarily for military, aerospace, and defense electronics, where devices must operate under extreme environmental conditions. These standards specify rigorous environmental and mechanical testing procedures, including vibration, thermal shock, and humidity testing, to evaluate the reliability of packaging systems in harsh conditions.

For example, MIL-STD-883 outlines methods for testing the environmental durability and mechanical strength of microelectronic components. It defines protocols for shock testing, moisture resistance, and corrosion prevention, ensuring that packages used in space missions, avionics, and military equipment can withstand extreme temperatures, mechanical impacts, and exposure to moisture or chemicals.

Testing under MIL-STD guidelines is essential for validating high-reliability systems, where any failure can have catastrophic consequences. Devices used in aerospace and defense applications undergo accelerated life testing and destructive testing to confirm that they meet the stringent performance requirements outlined by MIL-STD.

8.6.3   IPC Standards

The IPC (Institute for Printed Circuits) defines a comprehensive set of standards for the inspection, testing, and reliability assessment of PCBs, flexible circuits, and solder joints. IPC standards play a crucial role in ensuring that electronic assemblies meet quality benchmarks throughout the production cycle.

IPC-6012 sets the guidelines for rigid PCB fabrication, while IPC-A-610 provides visual criteria for solder joint inspection, helping manufacturers detect early defects. For flexible and hybrid electronics, IPC-2221 outlines design rules that ensure packaging systems maintain reliability even under bending or mechanical stress. These standards are essential for validating the performance of wearable devices, IoT sensors, and automotive electronics, where flexible circuits are widely used.

Compliance with IPC standards ensures that packages maintain consistent electrical performance, mechanical stability, and environmental durability, reducing the likelihood of failures during assembly or field use. Adhering to these guidelines also facilitates supply chain integration, enabling smooth collaboration between manufacturers and component suppliers.

8.6.4 Compliance Requirements and Certifications

Meeting industry standards is a critical step in achieving regulatory approval and building customer trust. In highly regulated industries such as automotive, healthcare, and aerospace, devices must pass multiple levels of testing and certification before they can be deployed. For example, ISO 13485 certification is required for medical devices, ensuring that electronic packaging systems used in healthcare meet safety and reliability standards. Similarly, automotive components must comply with AEC-Q100 testing guidelines to validate their performance under extreme operational conditions.

In addition to meeting formal compliance requirements, certifications from independent bodies—such as UL (Underwriters Laboratories) or TÜV Rheinland—offer further assurance to customers and stakeholders. These certifications validate that the product has undergone rigorous testing and meets international safety and quality benchmarks.

By adhering to established standards and obtaining certifications, manufacturers not only reduce the risk of product failures but also gain a competitive advantage in the market. Compliance demonstrates a commitment to quality and reliability, fostering trust among customers, partners, and regulators.

8.7   Reliability Modeling and Prediction

Reliability modeling and prediction play an essential role in evaluating the performance, lifespan, and failure risks of advanced packaging systems. As devices grow more complex and integrate multiple components, manufacturers rely on predictive models to estimate how materials and interconnects behave under operational stress over time. These models provide valuable insights into failure mechanisms, guiding engineers to optimize designs and improve reliability. Below are key reliability models and metrics used in advanced packaging, along with their practical applications.

8.7.1 Arrhenius Model

The Arrhenius model predicts the effect of temperature on the failure rate of materials and components. Based on the principle that chemical reactions accelerate at higher temperatures, this model estimates the rate at which materials degrade when exposed to elevated temperatures over time. It is particularly relevant for packaging systems subjected to thermal cycling or continuous operation at high temperatures, such as AI accelerators, automotive modules, and power electronics.

The model calculates the activation energy (Ea) required for a failure mechanism—such as corrosion, oxidation, or electromigration—to occur. Using this data, engineers can predict time-to-failure at different temperatures, helping them develop thermal management strategies to extend product life.

In practice, the Arrhenius model is used during accelerated stress tests, where devices are exposed to high temperatures to identify weak points. The results allow engineers to make material selections and design improvements that minimize the risk of temperature-induced failures in the field.

8.7.2   Coffin-Manson Model

The Coffin-Manson model estimates the fatigue life of materials subjected to thermal cycling. It is especially useful for predicting solder joint failures, which occur when materials expand and contract repeatedly under changing temperatures. Packaging systems that integrate multiple dies, substrates, and interconnects are vulnerable to fatigue due to differences in the CTE between materials.

By analyzing the strain induced during thermal cycling, the Coffin-Manson model predicts how long a material or joint will last before it develops cracks or fractures. The model plays a critical role in the design validation of products used in high-stress environments, such as automotive electronics, aerospace systems, and industrial machinery.

Testing protocols often involve accelerated thermal cycling, where devices are exposed to rapid temperature fluctuations to replicate years of real-world use in a compressed timeframe. The insights from these tests guide engineers in improving solder formulations and interconnect geometries, ensuring packaging durability under harsh conditions.

8.7.3   MTBF (Mean Time Between Failures)

MTBF is a reliability metric used to estimate the average lifespan of a device or system. It represents the expected time between two consecutive failures during normal operation. While MTBF does not predict when a specific failure will occur, it provides a useful benchmark for evaluating the overall reliability of complex systems.

This metric is widely applied in telecommunications, healthcare devices, and industrial equipment, where downtime must be minimized. For example, MTBF values help manufacturers determine maintenance schedules for critical systems and warranty periods for consumer products.

MTBF is typically calculated using failure rate data collected from accelerated life testing, giving manufacturers an estimate of how often repairs or replacements will be needed during the device's operational life. High MTBF values indicate better reliability and are a key selling point for products in mission-critical applications.

8.7.4   FIT (Failures in Time) Rate

The FIT rate measures the failure rate per billion hours of operation, offering a granular view of long-term reliability. FIT provides insight into how frequently failures are expected to occur over extended periods, making it an essential metric for devices used in long-term deployments, such as space probes, telecommunications infrastructure, and medical implants.

FIT is calculated from real-world field data or accelerated life tests and helps manufacturers identify which components or processes contribute most to the failure rate. This information is particularly valuable for system-level reliability modeling, enabling manufacturers to fine-tune designs and improve manufacturing processes. A low FIT rate reflects a highly reliable system, giving customers confidence in the product's performance over its intended lifespan.

8.7.5   Accelerated Life Testing

Accelerated life testing compresses the time required to predict field performance by subjecting products to high-stress conditions. This method helps engineers identify potential failure points early in the development process, allowing them to make design changes before mass production. Accelerated tests simulate extreme environmental conditions, such as high temperatures, humidity, mechanical vibrations, and voltage stress, replicating years of real-world operation in just weeks or months.

Common forms of accelerated testing include highly accelerated life testing (HALT) and highly accelerated stress screening (HASS). These tests reveal vulnerabilities that might not appear under normal operating conditions but could lead to field failures over time. For example, thermal cycling tests identify solder joint fatigue, while vibration tests highlight weak interconnects.

The results of accelerated life testing feed into reliability models such as the Arrhenius and Coffin-Manson models, providing data that allows manufacturers to predict product lifespan and optimize designs. By addressing potential failure points early, accelerated testing ensures that packaging systems meet reliability standards and perform consistently across their entire operational life.

8.8   Future Trends in Packaging Reliability and Testing

As advanced packaging technologies continue to evolve, the need for more sophisticated reliability testing methods grows. Emerging trends focus on new materials, AI-driven testing approaches, miniaturization challenges, and real-time monitoring. These innovations aim to enhance performance, durability, and predictive maintenance capabilities, ensuring that packaging systems remain robust under increasingly complex conditions. Below, we explore the key future trends shaping the field of reliability and testing.

8.8.1   AI-Powered Testing Systems

AI and machine learning (ML) are revolutionizing reliability testing by enabling automated defect detection and predictive maintenance. AI-powered systems can analyze large volumes of test data in real-time, identifying patterns and trends that might indicate future failures. By leveraging these insights, manufacturers can implement proactive measures to address issues before they lead to downtime.

Machine learning algorithms are also improving the accuracy of failure prediction models. For instance, AI-driven analyses can predict the lifespan of solder joints, interconnects, and adhesives more precisely, allowing for better design decisions and warranty planning. Additionally, automated optical inspection (AOI) systems powered by AI are being used to detect surface-level defects in large production batches, significantly enhancing efficiency and reducing inspection times.

8.8.2   Real-Time Monitoring and Digital Twins

Real-time monitoring with embedded sensors is transforming how packaging systems are managed throughout their lifecycle. Sensors [20, 21] integrated into packaging components can collect data on temperature, humidity, mechanical stress, and electrical performance, providing continuous insights into system health. This real-time data supports predictive maintenance, reducing unexpected downtime and improving operational efficiency [22].

The concept of digital twins—virtual replicas of physical systems—further enhances reliability testing and monitoring. By creating a real-time digital model of a packaging system, engineers can simulate various environmental and operational conditions, identifying potential failure points before they occur in the real world. Digital twins also enable remote diagnostics and optimization, helping manufacturers fine-tune their products throughout their lifecycle.

This approach is particularly beneficial in aerospace, automotive, and industrial applications, where reliability is critical, and downtime can have significant financial and safety implications. By combining real-time monitoring with AI-powered analytics, manufacturers can ensure that packaging systems meet performance standards even under the most demanding conditions.

Conclusion:

This chapter explored the future trends shaping packaging reliability and testing, emphasizing innovations in materials, AI-powered systems, and real-time monitoring technologies. Advancements such as graphene interconnects, self-healing polymers, and solid-state thermal interface materials promise enhanced durability and performance. As devices continue to miniaturize, new testing methodologies and micro-level simulations will be essential to validate ultra-dense packages. AI algorithms and digital twins are revolutionizing predictive maintenance, reducing downtime, and optimizing system performance. Together, these developments are paving the way for more resilient, reliable packaging solutions, ensuring that advanced electronics meet the growing demands of modern applications across diverse industries.

References

1. Guan, Shukai, et al. "A reliability assessment methodology of system-in-package based virtual qualification." *Microelectronics Reliability* 150 (2023): 115212.
2. Varshney, Nitin, et al. "Fault-marking: defect-pattern leveraged inherent fingerprinting of advanced IC package with thermoreflectance imaging." In *Infrared Sensors, Devices, and Applications XIV*, vol. 13145. SPIE, 2024.
3. Zhao, Shufeng, and Xingshou Pang. "Investigation of delamination control in plastic package." *Microelectronics Reliability* 49.3 (2009): 350–356.
4. Craig, Patrick, et al. "Multi-modal printed circuit board netlist extraction with x-ray and optical imaging." In *Developments in X-Ray Tomography XV*, vol. 13152. SPIE, 2024.
5. Ghosh, Shajib, et al. "A Framework for Overcoming Resolution and Sensitivity Limits in 7nm Node Technology Inspection via Automated Imaging and Analysis." (2024): ozae044-171.
6. Varshney, Nitin, et al. "Challenges and opportunities in non-destructive characterization of stacked IC packaging: insights from SAM and 3D x-ray analysis." In *Developments in X-Ray Tomography XV*, vol. 13152 (2024): 92–99.
7. Bender, Emmanuel, Joseph B. Bernstein, and Duane S. Boning. "Modern Trends in Microelectronics Packaging Reliability Testing." *Micromachines* 15.3 (2024): 398.
8. Park, C. E., et al. "Evaluation of epoxy underfill materials for use in the 'chip-on-board' method of packaging silicon integrated circuits." In *Polymer Surfaces and Interfaces: Characterization, Modification and Application*. CRC Press, 2023, pp. 361–374.
9. Noor, Rouhan, et al. "US microelectronics packaging ecosystem: Challenges and opportunities." arXiv preprint arXiv:2310.11651 (2023).
10. Supramaniam, Saraswathy, et al. "Moisture Content and Early Corrosion Detection of Cu Wire Bonding in a Semiconductor Package." *Journal of Failure Analysis and Prevention* 23.6 (2023): 2362–2369.
11. Li, Tengyu, et al. "Polymer-based nanocomposites in semiconductor packaging." *IET Nanodielectrics* 6.3 (2023): 147–158.
12. Ye, J. J., Wan-Ping Lien, and San Chen. "A Role of N2 and O2 Gasses in Post Etch Treatment (PET) for Removing Fluorocarbon Based By-Product Outgassing in DRAM Memory." In *2023 34th Annual SEMI Advanced Semiconductor Manufacturing Conference (ASMC)*. IEEE, 2023.

13. Varshney, Nitin, et al. "Challenges and opportunities in non-destructive characterization of stacked IC packaging: insights from SAM and 3D x-ray analysis." In *Developments in X-Ray Tomography XV*, vol. 13152 (2024): 92–99.
14. Shafkat, M., et al. "Assessing Compatibility of Advanced IC Packages to X-ray Based Physical Inspection." *EDFA Technical Articles* 26.3 (2024): 14–24.
15. Akiyama, Jotaro, Masanobu Naeshiro, and Masazumi Amagai. "A study of hot spot in silicon device for stacked die packages." In *2005 International Symposium on Electronics Materials and Packaging*. IEEE, 2005.
16. Xu, Jiefeng, et al. "An assessment of electromigration in 2.5 D packaging." In *2019 IEEE 69th Electronic Components and Technology Conference (ECTC)*. IEEE, 2019.
17. Craig, Patrick, et al. "Terahertz time-domain spectroscopy (THz-TDS) fingerprinting for integrated circuit (IC) identification in tracking and tracing applications." In *Terahertz Emitters, Receivers, and Applications XV*, vol. 13141. SPIE, 2024.
18. Khan, M. Shafkat M., et al. "Exploring advanced packaging technologies for reverse engineering a system-in-package (sip)." *IEEE Transactions on Components, Packaging and Manufacturing Technology* (2023).
19. Asadizanjani, Navid, Chengjie Xi, and Mark M. Tehranipoor. *Materials for Electronics Security and Assurance*. Elsevier, 2024.
20. Kottur, H. R., et al. "Enhancing the MEMS Gyroscope Physical Assurance Using Quantum Sensing." In *2024 IEEE Research and Applications of Photonics in Defense Conference (RAPID)*, Miramar Beach, FL, USA, 2024, pp. 1–2. https://doi.org/10.1109/RAPID60772.2024.10647048.
21. Biswas, L. K., et al. "Silicon Photonics Under Siege: Unveiling Security Vulnerabilities Against SAW." In *2024 IEEE Research and Applications of Photonics in Defense Conference (RAPID)*, Miramar Beach, FL, USA, 2024, pp. 1–2. https://doi.org/10.1109/RAPID60772.2024.10646987.
22. Khan, M., et al. "Assessing Compatibility of Advanced IC Packages to X-ray Based Physical Inspection." *Electronic Device Failure Analysis* 26.3 (2024).

Quantum Computing, Wearables, and Next-Gen IC Packaging

9

9.1 Foundations of the Quantum and Wearable Revolution

The digital landscape is evolving at an unprecedented pace, driven by breakthroughs in computing technologies and the growing ubiquity of smart devices. Two areas leading this transformation are quantum computing and wearable technologies. These domains address distinct challenges but share a common goal: to push the boundaries of what is technologically possible. Quantum computing [1] promises to solve problems beyond the reach of classical systems, while wearable devices revolutionize personal health, fitness, and communication. The intersection of these technologies with advanced semiconductor solutions signals a new era of innovation.

Quantum computing, although still in its nascent stages, is already showing enormous potential. Traditional computers, bound by binary logic, struggle to simulate complex quantum phenomena, optimize large datasets, or tackle cryptographic challenges efficiently [2]. Quantum systems leverage qubits [3] that exploit superposition and entanglement [4], enabling them to perform multiple operations simultaneously. Applications range from drug discovery and AI model optimization to climate modeling and financial forecasting [5]. However, achieving scalable, fault-tolerant quantum computers demands breakthroughs not only in quantum algorithms but also in system architecture and environmental control.

In parallel, wearable technologies are shaping how people interact with digital ecosystems. What started as fitness trackers and smartwatches has evolved into implantable medical devices [6], augmented reality headsets [7], and wearable biosensors [8]. These devices offer real-time health monitoring, assist in managing chronic conditions, and improve accessibility for people with disabilities. The growing interest in remote healthcare and wellness technology has made wearables an integral part of the IoT. However, as users demand more features in smaller, more comfortable devices, engineers must find new ways to integrate computing, sensing, and communication modules efficiently.

N. Asadizanjani et al., *Introduction to Microelectronics Advanced Packaging Assurance*, Synthesis Lectures on Engineering, Science, and Technology, https://doi.org/10.1007/978-3-031-86102-4_9

Another significant development shaping the future of both quantum and wearable technologies is the shift in semiconductor design paradigms. The industry is moving beyond the classical focus on Moore's Law, which emphasized increasing transistor density, toward innovations in heterogeneous integration and miniaturization. Emerging trends such as interposerless chip design promise to reduce system complexity and latency. These new approaches will unlock higher efficiency in quantum operations and provide wearables with enhanced battery life and user comfort.

The confluence of quantum computing, wearable devices, and semiconductor breakthroughs represents a pivotal moment in technology. As these fields mature, they will transform industries ranging from healthcare and telecommunications to cryptography and artificial intelligence. Quantum systems will make it possible to perform computations that were previously infeasible, while wearables will empower individuals to manage their health more proactively and efficiently. Together, these technologies have the potential to shape a smarter, more connected future.

This chapter explores the cutting-edge advances in both quantum computing and wearable devices and introduces the next wave of IC packaging innovations that will enable these technologies to reach new heights. The subsequent sections will delve into the key challenges and breakthroughs across these fields, emphasizing the role of scalable architectures, emerging packaging techniques, and hybrid systems in meeting the demands of tomorrow's digital world.

9.2   Quantum Computing and Advanced Packaging Technologies

Quantum computing signifies a monumental leap beyond the capabilities of classical systems. It introduces novel approaches to problem-solving using the unique properties of quantum mechanics, such as superposition and entanglement. As qubits—quantum bits—operate in multiple states simultaneously, they unlock new possibilities in computing power, allowing quantum systems to perform calculations that would take classical supercomputers centuries to complete. The development of quantum computers is already creating ripples across industries like cryptography, molecular modeling, drug discovery, and AI. However, for these systems to become practical, advanced packaging solutions are critical in maintaining the delicate quantum states and ensuring stable, scalable operations.

9.2.1   The Rise of Quantum Computing

Quantum computing has garnered significant momentum over the past decade, with leading organizations like IBM, Google, and Intel achieving key breakthroughs. IBM's Q System and Google's Sycamore processor [9] demonstrate how quantum computers can achieve

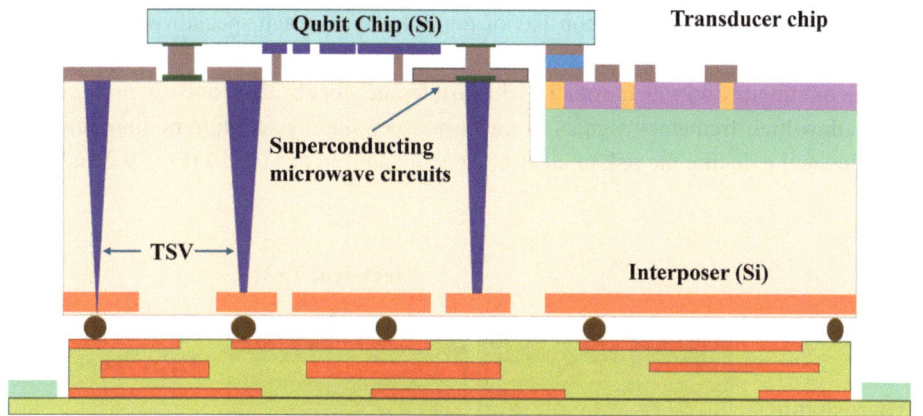

Fig. 9.1 Cross-sectional method for quantum chip interposer packaging

quantum supremacy performing tasks beyond the capability of classical supercomputers. For instance, Google's Sycamore executed [10] a complex computation in seconds, which would have taken a classical supercomputer 10,000 years to complete.

At the heart of quantum computing lies the challenge of maintaining qubit coherence [11]. Qubits are incredibly sensitive to thermal noise, electromagnetic interference, and environmental disruptions, which can cause them to lose their quantum state—a phenomenon known as decoherence [12]. To maintain qubit stability, quantum chips must operate at cryogenic temperatures (close to absolute zero) to reduce noise and preserve coherence. As quantum systems scale up to include more qubits and complex interconnections, thermal management, signal integrity, and scalable packaging solutions become paramount.

Moreover, quantum computing devices rely on microwave frequencies and optical components to perform computations, adding another layer of complexity. Packaging solutions must integrate optical and electrical signals seamlessly to ensure low-latency operations across different frequencies (Fig. 9.1).

9.2.2 3D Integration for Quantum Chips

The transition from traditional 2D planar designs to 3D integrated architectures is transforming quantum computing. In quantum systems, speed and precision are critical, making the reduction of signal path lengths a priority. 3D integration achieves this by stacking qubit arrays [13], control electronics, and readout circuits vertically, bringing them closer to each other and minimizing latency. This design not only enhances signal transmission but also helps avoid interference between components that would otherwise need to communicate across a planar surface.

A typical 3D quantum chip consists of multiple layers, each specialized for a distinct function—such as qubit control, readout, and error correction. Each layer must maintain precise communication with the others, with no tolerance for electrical noise or timing delays. TSVs allow high-frequency signals to travel between the layers while minimizing signal distortion and reducing the risk of electromagnetic interference (EMI) (Figs. 9.2 and 9.3).

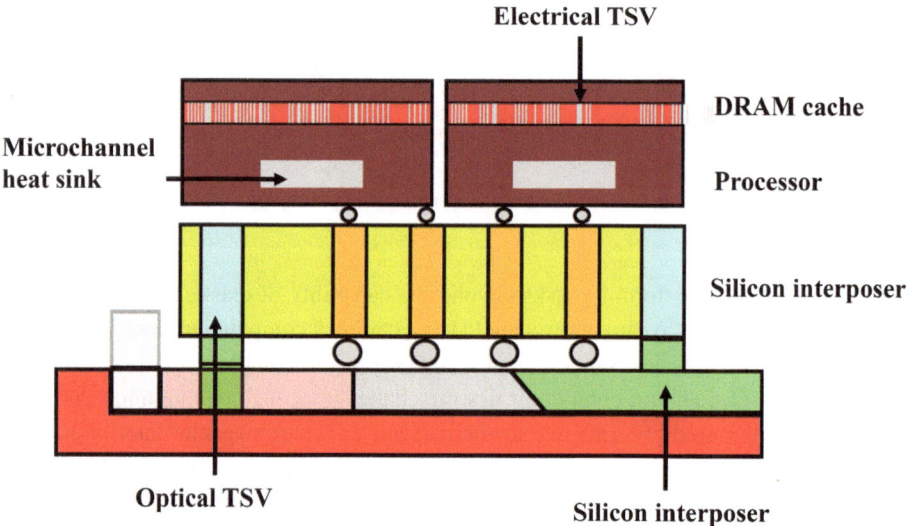

Fig. 9.2 Silicon interposer with electrical and optical TSVs in advanced packaging structures

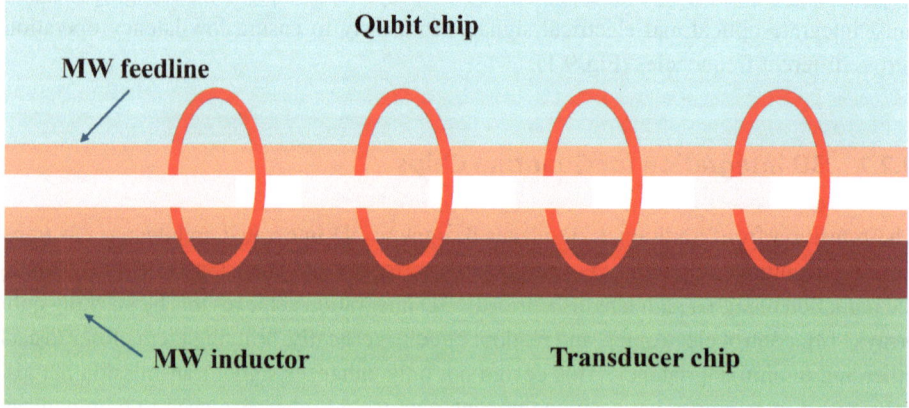

Fig. 9.3 Coupling techniques in quantum chip interposer packaging

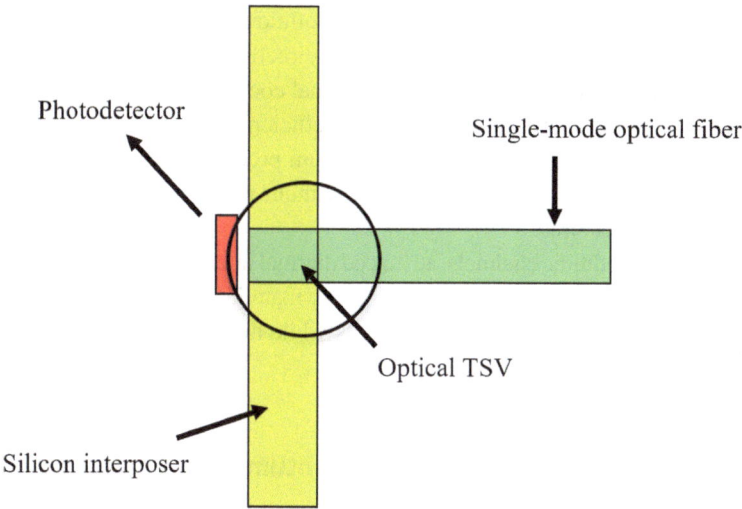

Fig. 9.4 Example of an optical TSV for optical interconnects in high-performance quantum operations

9.2.3   TSVs: Unlocking High-Performance Quantum Operations

TSVs are crucial to the efficient performance of quantum operations. In quantum systems that use superconducting qubits [14], control signals must be transmitted at microwave frequencies. Any delay or signal degradation compromises the fidelity of qubit states. TSVs provide a direct vertical path for signal flow between control circuits and qubit layers, eliminating the need for longer horizontal interconnects, which would introduce signal degradation and increase latency (Fig. 9.4).

Beyond mere signal routing, TSVs also contribute to the mechanical stability of the 3D stack by physically bonding the layers. This bonding ensures the precise alignment of qubit arrays with their corresponding control units, enabling fast, synchronized operations without clock drift. TSV-based designs offer a clear advantage for applications that demand high-frequency communication, such as quantum error correction protocols [15], where real-time feedback loops are essential.

9.2.4   Thermal Management in 3D Quantum Chips

A key challenge with 3D stacking is heat dissipation. While TSVs improve communication and latency, stacking multiple active layers generates heat, which can interfere with qubit coherence. Qubits, especially superconducting qubits, operate at cryogenic temperatures [16], and even a minor temperature increase can disrupt operations.

To address this, researchers are developing microfluidic cooling channels [17] within the 3D stack. These channels circulate coolant fluids directly through the chip structure, providing localized heat management. Unlike traditional cooling methods that rely on external refrigerators, integrated microfluidic systems can efficiently dissipate heat without disturbing the delicate thermal environment of the quantum processor. This solution ensures that high-frequency control signals are transmitted without performance degradation, even in densely packed architectures.

In addition to microfluidic channels, advanced thermal interface materials (TIMs) are used between layers to enhance heat transfer [18]. These materials allow heat to flow away from critical components like qubits and control units while maintaining the structural integrity of the stack.

9.2.5   Benefits of 3D Integration for Quantum Error Correction

Quantum error correction is one of the biggest hurdles to scaling quantum computers. As the number of qubits increases, the need for efficient error correction becomes more pressing. 3D architectures facilitate faster communication between qubit arrays and error-correcting circuits by stacking them directly on top of each other. This minimizes the delay in detecting and correcting errors, which is essential for maintaining qubit coherence.

Additionally, 3D designs enable the co-location of qubits and classical control circuits, which significantly improves system scalability. For instance, ancillary qubits used for error detection can be placed on a separate layer, reducing the computational burden on the primary qubits. This arrangement allows the system to scale efficiently without compromising qubit fidelity.

9.2.6   Intel's 3D Quantum Prototypes: A Case Study

Intel's quantum research team has demonstrated the potential of 3D integration for quantum processors through their prototype chips. These chips employ TSVs and copper pillar bonding to vertically stack qubits, control circuits, and readout layers. The design ensures high-speed communication between layers, with reduced latency and improved synchronization across the system.

Intel's prototype also incorporates adaptive thermal management solutions to ensure stable cryogenic operation. This includes integrating low-power cryogenic control electronics in the lower layers of the stack, minimizing the heat generated near the qubits. Such a design prevents thermal interference and extends the operational lifetime of qubit states, enabling the system to run more complex algorithms reliably.

9.3 Advanced Packaging for Wearable Devices

Wearable technologies sit at the intersection of microelectronics, healthcare, and consumer electronics, delivering real-time monitoring and enhanced interactivity. Whether in smartwatches, fitness trackers, AR glasses [19], or implantable medical sensors, packaging innovations are critical to ensure these devices are compact, energy-efficient, flexible, and durable. As the demand for seamless, always-on devices grows, engineers focus on new materials and system architectures to meet these requirements. In this section, we explore the specific challenges and advancements that define modern wearable technology.

9.3.1 Key Requirements for Wearables: Compact, Biocompatible, and Energy-Efficient

Wearables need to balance comfort, performance, and reliability, and this balance starts with miniaturization. Devices like smartwatches, fitness trackers, and earbuds require tight integration of multiple components into thin, lightweight assemblies. However, shrinking form factors introduces challenges, such as managing heat and optimizing power consumption. To address these, designers use thin-film circuits, low-power microcontrollers, and lightweight sensor modules. Additionally, thermal management must be handled delicately, as wearable electronics in contact with the skin should not overheat or cause discomfort. Phase-change materials (PCMs) are employed in high-performance [20] smartwatches to absorb sudden heat spikes during peak usage, improving user comfort without increasing the device's size.

Biocompatibility is another crucial requirement, especially for implantable medical devices such as glucose monitors, pacemakers, and smart contact lenses. These devices must use materials that do not trigger allergic reactions or cause infections. Silicone elastomers and Parylene coatings [21] are popular choices for their non-toxic, inert properties. In hydrogel-based ECG patches, the soft hydrogel ensures continuous contact with the skin, reducing motion artifacts that can affect data accuracy. For implantable devices, titanium shells or bio-inert polymers are used to avoid immune system rejection.

Power management is essential in wearables, as users expect devices to run for extended periods without frequent recharging. Ultra-low-power processors (ULPs) [22], often paired with dynamic power management systems, ensure efficient operation. Additionally, wireless charging and near-field communication (NFC) [23] modules enable convenient power delivery without the need for physical ports, further reducing device size.

9.3.2 Flexible and Bendable Packaging for Wearables

Wearable electronics must adapt to the movements of the human body, which makes flexible and stretchable materials essential. Traditional rigid PCBs are being replaced with polyimide films, liquid crystal polymers (LCPs) [24], and stretchable elastomers. These materials allow the devices to conform to the body, ensuring comfort and durability during everyday use.

Polyimide substrates are widely used in thin-film sensors for fitness trackers due to their thermal stability and mechanical flexibility. Liquid crystal polymers (LCPs) are preferred for high-frequency antennas embedded in wireless earbuds, as they provide excellent electrical properties with low moisture absorption. Stretchable elastomers, such as PDMS (polydimethylsiloxane), are used in electronic skin (e-skin) patches [25], which can monitor health metrics like hydration and body temperature. These elastomers allow sensors to bend and stretch without breaking, making them ideal for sports and medical wearables.

In addition to flexible substrates, wearables increasingly rely on MEMS for precise motion tracking and health monitoring. MEMS accelerometers, gyroscopes [26], and pressure sensors allow fitness trackers to measure physical activity, posture, and environmental conditions. For example, 3-axis accelerometers detect motion and steps in fitness bands, while gyroscopes improve orientation tracking in virtual reality (VR) headsets. Barometers embedded in smartwatches provide altitude data, enhancing outdoor tracking applications such as hiking and running (Fig. 9.5).

9.3.3   Energy Harvesting and Self-Powered Systems

To reduce dependency on external charging, wearable devices are increasingly integrating energy harvesting technologies. These systems capture ambient energy from body heat, motion, or light to generate power, enabling continuous operation.

Thermoelectric generators (TEGs) leverage the temperature difference between the wearer's skin and the surrounding environment to produce electricity [27]. This approach is particularly useful for fitness trackers and temperature-monitoring patches, which remain in direct contact with the skin. For example, a wrist-worn fitness band with TEGs can generate enough power to operate continuously without requiring daily charging.

Piezoelectric materials [28] offer another avenue for energy harvesting, converting mechanical stress into electrical energy. Wearable devices embedded with piezoelectric sensors can generate power from activities such as walking, running, or joint movement. A knee brace equipped with piezoelectric elements can harvest energy from joint flexion to power sensors that monitor the joint's health in real time.

Photovoltaic (solar) cells [29] integrated into smartwatches or wearable fabrics enable light-based energy harvesting. Thin-film solar modules can be embedded directly into the display of a smartwatch, capturing ambient light to supplement the battery. These self-powered systems reduce the need for frequent charging, enhancing the usability and lifespan of wearable devices.

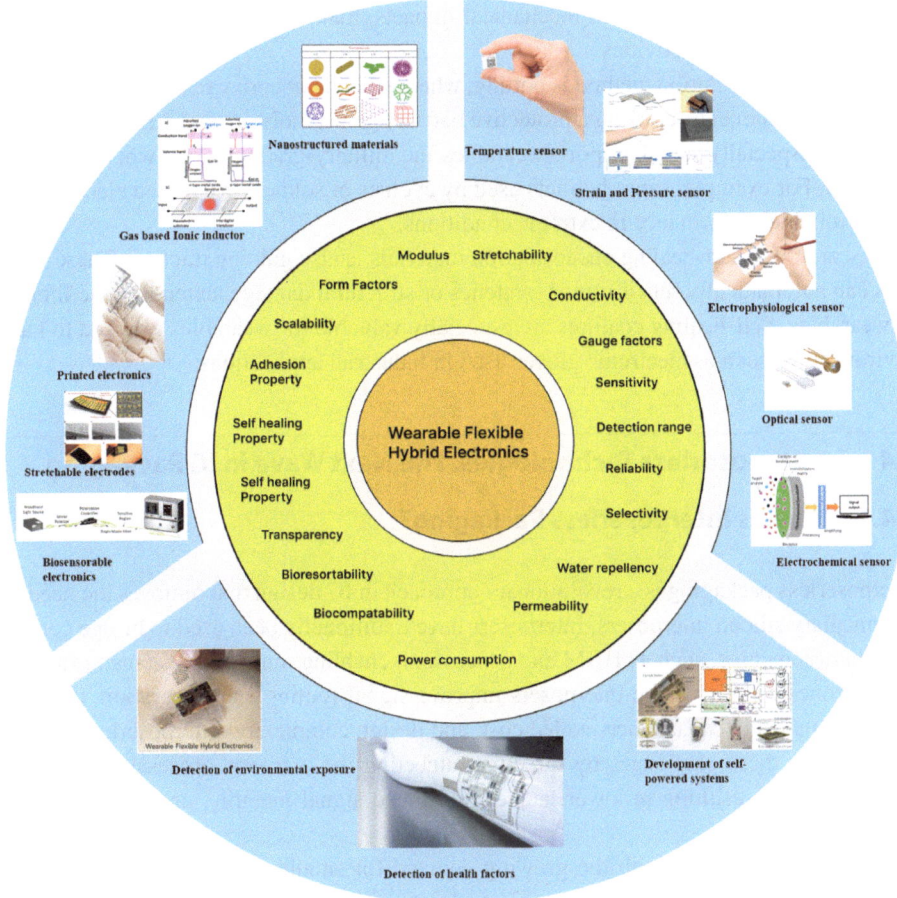

Fig. 9.5 Next-generation flexible and wearable electronics

9.3.4 Encapsulation for Rugged Wearables

Wearables, especially those designed for outdoor and sports applications, must withstand moisture, dust, impacts, and temperature fluctuations. Advanced encapsulation techniques ensure that electronic components are protected from environmental damage, extending the device's durability.

Thermoplastic Polyether Ester Elastomer (TPEE) is a widely used encapsulation [30] material due to its flexibility, waterproofing properties, and mechanical strength. TPEE housings seal microcontrollers and sensors within smartwatches, ensuring they remain operational even during swimming or high-humidity conditions. This material also provides resistance against scratches and mechanical impacts, making it suitable for rugged wearables used in outdoor activities.

Another encapsulation method is potting, where silicone or epoxy resins are poured over electronic components, creating a protective barrier against moisture and dust. Potting techniques are especially useful in sports wearables and military-grade devices, where durability is critical. For example, smart helmets used by cyclists or soldiers use potted electronics to ensure reliable performance in extreme conditions.

Research into self-healing encapsulation materials is also gaining traction. These materials can automatically repair minor scratches or structural damage, extending the lifespan of wearables. Self-healing coatings are especially valuable for wearables exposed to harsh environments, such as electronic gloves used in industrial applications.

9.4   Interposerless Technologies: The Next Wave in IC Packaging

9.4.1   What is Interposerless Packaging?

Interposerless packaging is a revolutionary approach in IC design that removes the need for intermediary silicon interposers. Interposers have traditionally been used to bridge connections between chips, such as HBM and processors, enabling multi-die systems to function as one cohesive unit. While interposers improve signal routing and integration, they also introduce parasitic capacitance, added cost, and design complexity. Interposerless packaging bypasses these challenges by directly connecting the memory, processing units, and functional dies, resulting in lower latency, enhanced signal integrity, and reduced power consumption.

This approach aligns with the growing demand for smaller, faster, and more efficient systems—especially in domains like AI accelerators, 5G networks, and quantum computing. By eliminating the interposer, data paths become shorter, minimizing signal distortion and delays, which are critical for high-speed, low-power applications. Moreover, interposerless packaging reduces manufacturing costs and thermal resistance, facilitating better heat management and easier system scaling.

9.4.2   Applications of Interposerless Packaging

The removal of interposers opens up new opportunities across several high-performance domains. As AI, quantum computing, and 5G networks demand lower latency and higher throughput, interposerless designs become increasingly appealing. Below, we explore the impact of interposerless packaging in key applications.

9.4.2.1 Interposerless Quantum Chips

Quantum computing systems, especially those using superconducting qubits and trapped-ion qubits, require ultra-high signal integrity. Even slight interference in the signal path can cause decoherence, rendering computations useless. In traditional quantum systems, silicon interposers are used to route microwave signals from qubits to control circuits. However, these interposers introduce parasitic capacitance, which degrades signal quality and limits coherence times.

By adopting interposerless packaging, quantum systems achieve direct interconnections between qubit arrays and control electronics, reducing latency and ensuring stable qubit coherence over longer periods. This approach also simplifies the design of cryogenic quantum processors, as fewer layers result in improved thermal efficiency. Removing the interposer enables tighter integration of optical or microwave control circuits, paving the way for quantum architectures that can scale more easily with less interference between components.

Example Use Case: A quantum processor [31] with interposerless connections achieves faster qubit operations by minimizing communication delays between qubit arrays and their readout systems. In multi-qubit systems, this approach reduces overhead, allowing for more complex calculations within shorter coherence windows.

9.4.3 Impact on AI Accelerators and 5G Networks

AI accelerators, such as Tensor Processing Units (TPUs) and GPUs, handle massive data loads that require rapid communication between memory and compute cores. Traditionally, these systems rely on 2.5D integration with interposers to connect multiple chips within a single package. However, the interposers introduce signal bottlenecks and complicate routing, particularly when dealing with high-frequency operations and MCMs [32].

In an interposerless architecture, the memory dies (such as HBM stacks) and compute cores are directly interconnected, shortening the data paths and improving data transfer speeds. This design not only boosts processing power but also reduces energy consumption—an essential requirement for data centers and edge AI devices. Additionally, the absence of interposers simplifies thermal management, enabling higher clock speeds without the risk of thermal throttling [33].

5G networks benefit significantly from interposerless packaging, as RF chips must process large amounts of data in real time with minimal delay. Direct interconnects between RF front-ends, antennas, and baseband processors enhance signal integrity and reduce latency. This is crucial for applications such as autonomous vehicles, smart cities, and industrial IoT, where even microseconds of delay can disrupt operations.

9.5   Advantages and Future Trends

Interposerless packaging offers several key advantages, making it a promising trend for next-generation ICs:

Lower Latency and Improved Signal Integrity: Direct connections between dies eliminate the need for routing through interposers, resulting in faster data transfer and minimal signal degradation. This is especially important for high-speed applications like AI inference engines and 5G modems, where even nanosecond-level delays can impact performance.

Reduced Power Consumption: In multi-chip systems, the absence of interposers minimizes power loss, making the design more energy-efficient. Lower power consumption is critical for battery-operated AI devices and edge computing nodes, where energy efficiency directly affects usability.

Simplified Manufacturing and Cost Reduction: Removing interposers from the design reduces both manufacturing complexity and costs. Fewer materials are required, and the overall assembly becomes more straightforward, resulting in faster production cycles and lower failure rates. This shift will be particularly beneficial for companies building large-scale AI infrastructure or quantum computers, where yield optimization is a priority.

Enhanced Scalability and Thermal Performance: With fewer layers to manage, interposerless designs offer improved thermal conductivity, enabling high-performance computing systems to maintain stability under heavy workloads. As systems scale to higher chip densities, better thermal performance ensures that devices can operate reliably at elevated frequencies without overheating.

9.6   Quantum Computing: Overcoming Decoherence and Ensuring Fault Tolerance

9.6.1   Quantum Decoherence: Preserving Qubit Stability

Quantum decoherence refers to the loss of a qubit's quantum state due to interaction with the environment, causing it to behave like a classical bit. Even the smallest temperature fluctuation, electromagnetic interference, or mechanical vibration can introduce noise, collapsing the superposition state of qubits. This sensitivity makes packaging for quantum processors extremely challenging, as they must operate in cryogenic environments to reduce thermal noise and maintain coherence.

To counter decoherence, quantum chips are housed in dilution refrigerators that cool components to millikelvin temperatures. However, even at these ultra-low temperatures, packaging materials can introduce EMI through residual impurities or insufficient shielding.

Quantum processors also require precise microwave control and signal isolation to prevent crosstalk between qubits. Packaging solutions must ensure that control circuits, qubit arrays, and readout systems are thermally and electrically isolated, which adds complexity to the overall architecture.

Emerging Solutions:

Superconducting shielding layers integrated into packaging designs help block electromagnetic interference. Researchers are also experimenting with diamond-based qubits, which show higher resilience to environmental noise than superconducting qubits. Low-vibration cryo-packaging [34] designs are being explored to minimize mechanical disturbances that could affect qubit coherence. Despite these advances, achieving long-lasting coherence remains a major hurdle, especially as quantum systems scale up to incorporate thousands or millions of qubits.

9.6.2 Fault Tolerance: Managing Errors in Quantum Operations

Quantum computing systems are inherently prone to errors due to noise and decoherence, necessitating robust error-correction mechanisms to maintain accuracy during computations. However, quantum error correction is far more complex than its classical counterpart, requiring multiple physical qubits to represent a single logical qubit. For example, a single logical qubit in a fault-tolerant quantum computer may require 50 to 100 physical qubits to correct errors effectively.

Current packaging designs struggle to accommodate the spatial and power requirements of these large qubit arrays while maintaining high fidelity. Furthermore, the control circuits needed to correct errors in real time introduce additional complexity. The control signals must be transmitted without introducing noise or delays that could interfere with qubit operations, which requires high-precision routing and synchronization across multiple layers of the package.

Potential Strategies:

Cryogenic control electronics, such as Intel's Horse Ridge [35], bring control circuits closer to qubits, reducing latency and noise.

Topological qubits, which are less prone to environmental disturbances, are being explored as a path toward scalable, fault-tolerant quantum systems.

Even with these strategies, scaling quantum computers to handle complex algorithms remains a significant challenge due to the limitations of existing error-correction codes [36] and packaging constraints.

9.7   Wearable Devices: Ensuring Biocompatibility and Environmental Reliability

9.7.1   Biocompatibility: Safe Interaction with the Human Body

Biocompatibility is a critical concern for medical wearables and implantable devices. Devices such as glucose monitors, pacemakers, and smart contact lenses must be designed with materials that are non-toxic, hypoallergenic, and safe for prolonged contact with the body. Packaging materials that do not meet these standards can cause skin irritation, infections, or immune responses, compromising both user safety and device performance.

To ensure biocompatibility, manufacturers rely on silicone elastomers, hydrogels, Parylene coatings, and titanium housings. These materials are not only biologically inert but also resistant to sweat, oils, and other bodily fluids. In implantable devices, the packaging must prevent corrosion and biofouling while allowing selective diffusion of molecules, such as glucose or oxygen, for real-time monitoring.

Challenges and Solutions:

Skin patches designed for continuous health monitoring need to maintain adhesiveness without irritating the skin, even during sweating or physical activity. Hydrogels, which provide both adhesion and flexibility, are frequently used but can degrade over time. Researchers are developing self-healing hydrogel coatings to extend device lifetimes.

Implantable electronics, such as pacemakers, must be hermetically sealed to prevent moisture ingress while remaining lightweight and non-invasive. Titanium housings provide an optimal balance between durability and biocompatibility for long-term implants.

Achieving a high degree of biocompatibility without compromising performance is an ongoing challenge, especially as wearables become more sophisticated and embedded with multi-sensor modules.

9.7.2   Environmental Reliability: Surviving Harsh Conditions

Wearable devices are often exposed to unpredictable environmental conditions, such as extreme temperatures, moisture, dust, and physical impacts. Fitness trackers, smartwatches, and AR glasses must remain operational in diverse settings—from hot, humid environments to cold, dry climates—without experiencing performance degradation. Ensuring waterproofing, dust resistance, and impact protection is essential, especially for outdoor and sports wearables.

Ingress Protection (IP) ratings are used to certify the level of environmental protection provided by wearable devices. For example, many smartwatches now feature IP68 certification, indicating resistance to dust and immersion in water up to a certain depth. However, maintaining such reliability requires advanced encapsulation techniques, including potting and conformal coatings, which seal sensitive components from environmental hazards.

Encapsulation Techniques:

Potting involves filling the internal spaces of the device with silicone or epoxy resin, providing protection against moisture and mechanical stress. Potting is commonly used in rugged wearables designed for extreme sports and military applications.

Conformal coatings create a thin, flexible barrier around electronic components, protecting them from corrosion and dust without adding significant weight. These coatings are especially useful in smart fabrics and wearable sensors, which need to remain lightweight and breathable.

Another environmental challenge is ensuring that wearables remain functional across wide temperature ranges. Components such as lithium-ion batteries can degrade rapidly in extreme temperatures, reducing device lifespan. New developments in solid-state batteries and temperature-resistant polymers aim to improve reliability and extend the operational range of wearable electronics.

9.8   Research Directions and Future Opportunities

As quantum computing and wearable technologies evolve, the research landscape is shifting toward more advanced, scalable architectures, materials, and integration techniques. The future will see quantum systems achieve higher operational stability through innovative methods, while wearables move toward autonomous and AI-powered solutions. This section explores new research areas and opportunities in both fields, focusing on novel developments beyond the challenges discussed earlier.

9.8.1   Quantum Computing Research Directions

9.8.1.1 Fault-Tolerant Quantum Systems and Novel Qubit Architectures

Beyond traditional superconducting qubits, researchers are investigating topological qubits as a more robust approach to error correction. Topological qubits, based on exotic particles called anyons, offer inherent resistance to environmental noise by encoding quantum information in the topology of particle braids. This design significantly reduces the overhead required for error correction and opens the door to more practical, large-scale quantum computers.

Another promising area of research focuses on spin qubits in silicon, which leverage existing CMOS infrastructure. Spin qubits exhibit longer coherence times than superconducting qubits and can be manufactured using well-established silicon processes, making them highly scalable. By developing hybrid systems that combine topological and spin qubits, researchers aim to create fault-tolerant architectures with both stability and scalability.

Research Challenge: Creating quantum error correction algorithms optimized for topological qubits requires new mathematics and protocols. This ongoing research aims to reduce the resource overhead currently needed to implement error correction, making fault-tolerant quantum computing more viable.

9.8.1.2 Integrated Photonics and Quantum Communication Networks

In the future, integrated photonics will play a crucial role in building quantum networks, enabling long-distance quantum communication. Photonic qubits—encoded in individual photons—are immune to decoherence, making them ideal for transmitting quantum information across vast distances.

Current research is focused on integrating photonic components directly into quantum processors, allowing on-chip quantum communication between qubits without relying on electrical signals. This will enable distributed quantum computing, where multiple quantum cores communicate seamlessly to solve complex problems in parallel. Silicon photonics platforms are being developed to create quantum processors that use optical interconnects for data transmission at the speed of light.

Another critical research direction involves quantum repeaters, devices that extend the range of quantum communication networks by reducing photon loss. Entanglement swapping protocols [37] are being optimized to create robust, large-scale quantum networks, paving the way for quantum internet infrastructure.

Opportunity: Developing optical quantum routers that can dynamically route quantum signals across networks will enable the creation of scalable quantum data centers [38].

9.8.2   Wearables Research Directions

9.8.2.1 Flexible Hybrid Electronics (FHE) and Advanced Biointerfaces

FHE represent the next frontier in wearables, combining rigid silicon-based components with flexible substrates. Researchers are exploring ultra-thin stretchable batteries, flexible antenna arrays, and bendable microcontrollers to build electronics that conform perfectly to the human body. These systems are integrated with biosensors capable of monitoring multiple physiological parameters simultaneously, such as blood glucose, hydration levels, and cortisol concentrations.

New materials such as graphene-based conductors are being investigated for use in electronic tattoos—ultrathin devices that adhere to the skin like temporary tattoos. These tattoos provide continuous health monitoring without discomfort and can interface with smartphones via Bluetooth. Additionally, 3D-printed biosensors are enabling custom wearable solutions tailored to individual patients, improving diagnostic precision.

Future Opportunity: Researchers are working on implantable FHE devices for neural interfaces, enabling direct communication between the brain and external devices. These systems hold the potential to restore motor function for individuals with paralysis and advance brain-computer interface technologies [39].

9.8.2.2 Autonomous Systems and AI-Driven Wearables

The future of wearables lies in autonomous systems that leverage AI-powered sensors to provide predictive and preventive healthcare. These devices will not only monitor physiological data but also use machine learning algorithms to predict health events, such as heart attacks or epileptic seizures, before they occur.

One area of research focuses on neuromorphic processors, which mimic the structure of biological neurons to process sensory data with ultra-low power consumption. These processors are integrated into wearables to perform real-time analysis without requiring constant cloud connectivity, ensuring faster response times. Wearables equipped with neuromorphic chips can learn and adapt to individual users, providing personalized recommendations for fitness, stress management, and sleep improvement.

Use Case Example: A smart ring with a neuromorphic AI chip could predict arrhythmias based on subtle changes in heart rate patterns, alerting the user and healthcare providers in advance.

Another research focus is the development of self-repairing AI models that allow wearables to detect and correct faults autonomously. These models ensure continuous operation by dynamically reallocating tasks to functional components in case of hardware failures or environmental interference.

Opportunity: As wearables become more autonomous, they will play a key role in remote patient monitoring systems, enabling continuous care for individuals with chronic conditions while reducing the burden on healthcare facilities.

Conclusion:

This chapter examined the intersection of quantum computing, wearable devices, and advanced IC packaging. Emerging technologies, such as interposerless architectures, flexible hybrid electronics, and integrated photonics, are shaping the next generation of computing and healthcare solutions. Innovations in cryogenic packaging, energy harvesting systems, and autonomous AI sensors are driving these fields forward, overcoming challenges of coherence, durability, and scalability. As interdisciplinary research accelerates, these advancements will unlock transformative applications across industries, positioning advanced packaging as the core enabler of future breakthroughs in high-performance computing and personalized health monitoring.

References

1. Sood, Vaishali, and Rishi Pal Chauhan. "Archives of quantum computing: research progress and challenges." *Archives of Computational Methods in Engineering* 31.1 (2024): 73–91.
2. Menezes, Alfred, and Douglas Stebila. "Challenges in cryptography." *IEEE Security & Privacy* 19.2 (2021): 70–73.
3. Chae, Eunmi, Joonhee Choi, and Junki Kim. "An elementary review on basic principles and developments of qubits for quantum computing." *Nano Convergence* 11.1 (2024): 11.
4. Forcer, Tim M., et al. "Superposition, entanglement and quantum computation." *Quantum Information and Computation* 2.2 (2002): 97–116.
5. Kristian, Agus, et al. "Application of AI in optimizing energy and resource management: Effectiveness of deep learning models." *International Transactions on Artificial Intelligence* 2.2 (2024): 99–105.
6. Mahmud, Sultan, et al. "Harnessing metamaterials for efficient wireless power transfer for implantable medical devices." *Bioelectronic Medicine* 10.1 (2024): 7.
7. Pérez-Pachón, Laura, et al. "Augmented reality headsets for surgical guidance: the impact of holographic model positions on user localisation accuracy." *Virtual Reality* 28.2 (2024): 68.
8. Wu, Zixiong, et al. "Interstitial fluid-based wearable biosensors for minimally invasive healthcare and biomedical applications." *Communications Materials* 5.1 (2024): 33.
9. AbuGhanem, Muhammad. "Google Quantum AI's Quest for Error-Corrected Quantum Computers." arXiv preprint arXiv:2410.00917 (2024).
10. Horner, Jack K., and John F. Symons. "What Have Google's Random Quantum Circuit Simulation Experiments Demonstrated about Quantum Supremacy?." In *Advances in Software Engineering, Education, and e-Learning: Proceedings from FECS'20, FCS'20, SERP'20, and EEE'20.* Springer International Publishing, 2021.
11. Chen, Yu, et al. "Qubit architecture with high coherence and fast tunable coupling." *Physical Review Letters* 113.22 (2014): 220502.
12. Brandt, Howard E. "Qubit devices and the issue of quantum decoherence." *Progress in Quantum Electronics* 22.5-6 (1999): 257–370.
13. Harrington, Patrick M., et al. "Synchronous detection of cosmic rays and correlated errors in superconducting qubit arrays." arXiv preprint arXiv:2402.03208 (2024).
14. McEwen, Matt, et al. "Resisting high-energy impact events through gap engineering in superconducting qubit arrays." arXiv preprint arXiv:2402.15644 (2024).
15. Postler, Lukas, et al. "Demonstration of fault-tolerant steane quantum error correction." *PRX Quantum* 5.3 (2024): 030326.
16. Stefanou, Georgios, and Charles Smith. "Calculation and design of GaAs quantum dot devices where the vibrational modes can be frozen out at cryogenic temperatures." *Semiconductor Science and Technology* (2024).

17. Boutsikakis, Athanasios, et al. "Glacierware: Hotspot-aware Microfluidic Cooling for High TDP Chips using Topology Optimization." arXiv preprint arXiv:2408.15024 (2024).

18. Cai, Wanwan, et al. "Aluminum/Graphene Thermal Interface Materials with Positive Temperature Dependence." *ACS Applied Materials & Interfaces* 16.26 (2024): 33993–34000.

19. Jang, Changwon, et al. "Waveguide holography for 3D augmented reality glasses." *Nature Communications* 15.1 (2024): 66.

20. Togun, Hussein, et al. "A critical review on phase change materials (PCM) based heat exchanger: different hybrid techniques for the enhancement." *Journal of Energy Storage* 79 (2024): 109840.

21. Pak, A. "Thermoplastic polymers for neural implantable interfaces." (2024).

22. Orlandi, Mattia, et al. "Real-Time Motor Unit Tracking from sEMG Signals with Adaptive ICA on a Parallel Ultra-Low Power Processor." *IEEE Transactions on Biomedical Circuits and Systems* (2024).

23. Liu, Yuanwei, et al. "Near-field communications: A comprehensive survey." arXiv preprint arXiv:2401.05900 (2024).

24. Lan, Ruochen, et al. "Adaptive liquid crystal polymers based on dynamic bonds: From fundamentals to functionalities." *Responsive Materials* 2.1 (2024): e20230030.

25. Tang, Desha, et al. "Inhibiting efflorescence of alkali-activated slag mortar by improving its hydrophobicity with methyl-terminated polydimethylsiloxane (PDMS): Evidences showing the remained PDMS." *Construction and Building Materials* 443 (2024): 137688.

26. H. R. Kottur, A. A. Khan, N. Varshney, L. K. Biswas, H. Dalir and N. Asadizanjani, "Enhancing the MEMS Gyroscope Physical Assurance Using Quantum Sensing," *2024 IEEE Research and Applications of Photonics in Defense Conference (RAPID)*, Miramar Beach, FL, USA, 2024, pp. 1–2, https://doi.org/10.1109/RAPID60772.2024.10647048.

27. He, Jifu, et al. "Advances in the applications of thermoelectric generators." *Applied Thermal Engineering* 236 (2024): 121813.

28. Ghemari, Zine, Salah Belkhiri, and Salah Saad. "A piezoelectric sensor with high accuracy and reduced measurement error." *Journal of Computational Electronics* 23.2 (2024): 448–455.

29. Li, Xin, et al. "Dimensional diversity (0D, 1D, 2D, 3D) in Perovskite solar cells: Exploring the potential of mix-dimensional integrations." *Journal of Materials Chemistry A* (2024).

30. Yang, Tao, et al. "Fabrication of Thermoplastic Poly (ether-ester) Elastomers with High Melting Temperature and Elasticity from Bio-based 2, 5-Furandicarboxylic Acid." *ACS Sustainable Resource Management* 1.7 (2024): 1520–1533.

31. Bluvstein, Dolev, et al. "Logical quantum processor based on reconfigurable atom arrays." *Nature* 626.7997 (2024): 58–65.

32. Lin, Qian, et al. "Thermal Characteristic Investigation for a Multichip Module Based on APDL." *International Journal of RF and Microwave Computer-Aided Engineering* 2024.1 (2024): 2028369.

33. Geb, David, et al. "Power Throttling in a 3D Integrated Circuit (IC) Dynamic Thermal Simulation." *Power* (2024).

34. Bhagabati, Purabi, Deepshikha Das, and Vimal Katiyar. "Bamboo-flour-filled cost-effective poly (ε-caprolactone) biocomposites: a potential contender for flexible cryo-packaging applications." *Materials Advances* 2.1 (2021): 280–291.

35. Mounier, Eric. "Quantum Technologies Are Speeding to Commercialization: Is the cat dead or alive or both? What is quantum?." *PhotonicsViews* 17.6 (2020): 27–29.

36. Steane, Andrew M. "Error correcting codes in quantum theory." *Physical Review Letters* 77.5 (1996): 793.

37. Dai, Wenhan, Anthony Rinaldi, and Don Towsley. "Entanglement swapping in quantum switches: Protocol design and stability analysis." arXiv preprint arXiv:2110.04116 (2021).

38. Martin, Michael James, et al. "Energy use in quantum data centers: Scaling the impact of computer architecture, qubit performance, size, and thermal parameters." *IEEE Transactions on Sustainable Computing* 7.4 (2022): 864–874.
39. Khan, Yasser, et al. "Flexible hybrid electronics: Direct interfacing of soft and hard electronics for wearable health monitoring." *Advanced Functional Materials* 26.47 (2016): 8764–8775.

Glossary

TSV Through-Silicon Via
RDL Redistribution Layer
FOWLP Fan-Out Wafer-Level Packaging
CVD Chemical Vapor Deposition
PVD Physical Vapor Deposition
DRIE Deep Reactive Ion Etching
WLP Wafer-Level Packaging
RIE Reactive Ion Etching
MOCVD Metal-Organic Chemical Vapor Deposition
PECVD Plasma-Enhanced Chemical Vapor Deposition
ALD Atomic Layer Deposition
IC Integrated Circuit
3DIC Three-Dimensional Integrated Circuit
MCMs Multi-Chip Modules
CTE Coefficient of Thermal Expansion
ECD Electrochemical Deposition
SIP System-in-Package
AIP Antenna-in-Package
HBM High-Bandwidth Memory
IOT Internet of Things
AI Artificial Intelligence
FHE Flexible Hybrid Electronics
EMI Electromagnetic Interference
RF Radio Frequency
GPUs Graphics Processing Units
CPUs Central Processing Units
TPUs Tensor Processing Units

© The Editor(s) (if applicable) and The Author(s), under exclusive license to Springer Nature Switzerland AG 2025
N. Asadizanjani et al., *Introduction to Microelectronics Advanced Packaging Assurance*,
Synthesis Lectures on Engineering, Science, and Technology,
https://doi.org/10.1007/978-3-031-86102-4

MEMS Microelectromechanical Systems
NEMS Nanoelectromechanical Systems
PDMS Polydimethylsiloxane
LCPs Liquid Crystal Polymers
PCBs Printed Circuit Boards
ULPs Ultra-Low-Power Processors
TIMs Thermal Interface Materials
HALT Highly Accelerated Life Testing
HASS Highly Accelerated Stress Screening
SEM Scanning Electron Microscopy
NDT Non-Destructive Testing
SMDs Surface-Mount Devices
TDR Time-Domain Reflectometry
EVs Electric Vehicles
DC Direct Current
HI Hybrid Integration
CMP Chemical Mechanical Planarization
STI Shallow Trench Isolation
HVM High-Volume Manufacturing
R&D Research and Development
QCMs Quartz Crystal Microbalances
AC Alternating Current
E-beam Electron Beam Evaporation
BEOL Back End of Line
ALE Atomic Layer Etching
HF Hydrofluoric Acid
FPCBs Flexible Printed Circuit Boards
KOH Potassium Hydroxide
HDPECVD High-Density Plasma Enhanced Chemical Vapor Deposition
CNTs Carbon Nanotubes
LPCVD Low-Pressure Chemical Vapor Deposition
LEDs Light Emitting Diodes
CoWoS Chip on Wafer on Substrate
InFO Integrated Fan-Out
HPC High-Performance Computing
EMIB Embedded Multi-die Interconnect Bridge
OSAT Outsourced Semiconductor Assembly and Test
DRAM Dynamic Random Access Memory
CMOS Complementary Metal-Oxide-Semiconductor
SoC System-on-Chip
C2 Chip-to-Chip Connections

C4 Controlled Collapse Chip Connection
I/O Inputs and Outputs
PUFs Physically Unclonable Functions
CAGR Compound Annual Growth Rate
IDMs Integrated Device Manufacturers
ATE Automated Test Equipment
CSP Chip-Scale Packaging
BGA Ball Grid Array
SMT Surface-Mount Technology